Iosias Jody (Ed.)

Channel Lake, Illinois

Iosias Jody (Ed.)

Channel Lake, Illinois

Census-designated place, Antioch Township, Lake County, Illinois

Cred Press

Imprint

Publisher:
Cred Press is a trademark of
International Book Market Service Ltd., 17 Rue Meldrum, Beau Bassin, 1713-01 Mauritius
Email: info@bookmarketservice.com
Website: www.bookmarketservice.com

Published in 2011

Printed in: U.S.A., U.K., Germany. This book was not produced in Mauritius.

ISBN: 978-613-7-40679-3

Contents

Articles

References

Channel Lake, Illinois

Channel Lake	
Unincorporated	
Country	United States
State	Illinois
County	Lake
Coordinates	42°28′40″N 88°8′55″W
Area	2.3 sq mi (6 km^2)
- land	2.0 sq mi (5 km^2)
- water	0.3 sq mi (1 km^2)
Population	1664 (*2010*)
Density	723 / sq mi (279 / km^2)
Timezone	CST (UTC-6)
- summer (DST)	CDT (UTC-5)

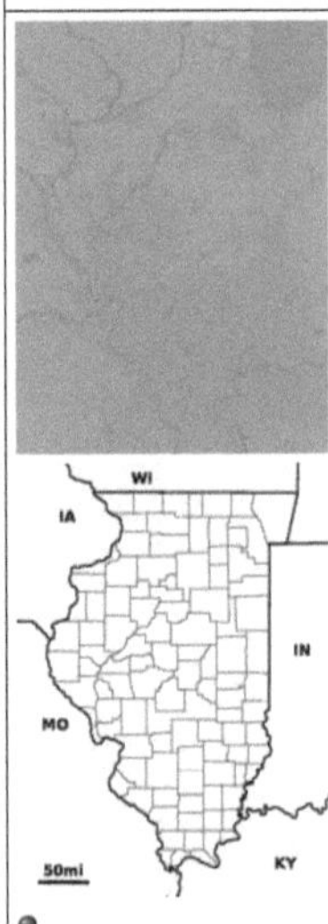

Location of Channel Lake within Illinois

Channel Lake is a census-designated place (CDP) in Antioch Township, Lake County, Illinois, United States. The population was 1,664 at the 2010 census.[1]

Geography

Channel Lake is located at 42°28′40″N 88°8′55″W.[2]

According to the United States Census Bureau, the CDP has a total area of 2.3 square miles (6.0 km^2), of which, 2.0 square miles (5.2 km^2) of it is land and 0.3 square miles (0.78 km^2) of it (14.85%) is water.

Demographics

As of the census[3] of 2000, there were 1,785 people, 696 households, and 473 families residing in the CDP. The population density was 916.1 people per square mile (353.4/km^2). There were 855 housing units at an average density of 438.8/sq mi (169.3/km^2). The racial makeup of the CDP was 96.13% White, 0.22% African American, 0.90% Native American, 0.73% Asian, 1.18% from other races, and 0.84% from two or more races. Hispanic or Latino of any race were 1.85% of the population.

There were 696 households out of which 32.5% had children under the age of 18 living with them, 52.4% were married couples living together, 8.9% had a female householder with no husband present, and 31.9% were non-families. 24.4% of all households were made up of individuals and 7.0% had someone living alone who was 65 years of age or older. The average household size was 2.56 and the average family size was 3.06.

In the CDP the population was spread out with 25.2% under the age of 18, 6.9% from 18 to 24, 31.8% from 25 to 44, 25.4% from 45 to 64, and 10.7% who were 65 years of age or older. The median age was 38 years. For every 100 females there were 112.2 males. For every 100 females age 18 and over, there were 112.2 males.

The median income for a household in the CDP was $51,384, and the median income for a family was $62,917. Males had a median income of $39,107 versus $30,050 for females. The per capita income for the CDP was $27,772. About 7.6% of families and 9.9% of the population were below the poverty line, including 19.6% of those under age 18 and 4.8% of those age 65 or over.

References

[1] "American FactFinder" (http://factfinder2.census.gov/main.html). U.S. Census Bureau. . Retrieved 9 March 2011.

[2] "US Gazetteer files: 2010, 2000, and 1990" (http://www.census.gov/geo/www/gazetteer/gazette.html). United States Census Bureau. 2011-02-12. . Retrieved 2011-04-23.

[3] "American FactFinder" (http://factfinder.census.gov). United States Census Bureau. . Retrieved 2008-01-31.

Census-designated place

A **census-designated place** (**CDP**) is a concentration of population identified by the United States Census Bureau for statistical purposes. CDPs are delineated for each decennial census as the statistical counterparts of incorporated places such as cities, towns and villages. CDPs are populated areas that lack separate municipal government, but which otherwise physically resemble incorporated places.

CDPs are delineated solely to provide data for settled concentrations of population that are identifiable by name but are not legally incorporated under the laws of the state in which they are located. They include small rural communities, colonias located along the U.S. border with Mexico, and unincorporated resort and retirement communities.[1] The boundaries of a CDP have no legal status.[2] Thus, they may not always correspond with the local understanding of the area or community with the same name. However, criteria established for the 2010 Census require that a CDP name "be one that is recognized and used in daily communication by the residents of the community" (not "a name developed solely for planning or other purposes") and recommend that a CDP's boundaries be mapped based on the geographic extent associated with residents' use of the place name.[1]

The U.S. Census bureau states that census-designated places are not considered incorporated places and that it includes census-designated places in Hawaii in its city population lists because that state has no incorporated cities.[3] In addition, census city lists from 2007 include Arlington County, Virginia's CDP in the list with the incorporated places.[4]

History

The Census Bureau reported data for some unincorporated places as early as the 1850 Census, though usage continued to develop through the 1890 Census, in which, for the first time, the Census mixed unincorporated places with incorporated places in its products with "town" or "village" as its label.[5] This made it confusing to determine which of the "towns" were or were not incorporated.[5]

The 1900 through 1930 Censuses did not report data for unincorporated places.[5]

For the 1940 Census, the Census Bureau compiled a separate report of unofficial, unincorporated communities of 500 or more people.[5] The Census Bureau officially defined this category as "unincorporated places" in the 1950 Census and used that term through the 1970 Census.[5] For the 1950 Census, these types of places were identified only outside "urbanized areas".[5] In 1960, the Census Bureau also identified unincorporated places inside urbanized areas (except in New England), but with a population of at least 10,000.[5] For the 1970 Census, the population threshold for "unincorporated places" in urbanized areas was reduced to 5,000.[5] For the 1980 Census, the designation was changed to "census-designated places"[5] and the designation was made available for places inside urbanized areas in New England.[5] For the 1990 Census, the population threshold for CDPs in urbanized areas was reduced to 2,500.[5] From 1950 through 1990, the Census Bureau specified other population requirements for unincorporated places or CDPs in Alaska, Puerto Rico, island areas, and Native American reservations. Minimum population criteria for CDPs were dropped with the 2000 Census.[1] [6]

The Census Bureau's Participant Statistical Areas Program (PSAP) allows designated participants to review and suggest modifications to the boundaries for CDPs.[7] The PSAP was to be offered to county and municipal planning agencies during 2008.

Effects of designation and examples

The boundaries of such places may be defined in cooperation with local or tribal officials, but are not fixed, and do not affect the status of local government or incorporation; the territories thus defined are strictly statistical entities. CDP boundaries may change from one census to the next to reflect changes in settlement patterns.[8] [5] Further, as statistical entities, the boundaries of the CDP may not correspond with local understanding of the area with the same name. Recognized communities may be divided into two or more CDPs while on the other hand, two or more communities may be combined into one CDP. A CDP may also cover the unincorporated part of a named community where the rest lies within an incorporated place.

By defining an area as a CDP, that locality then appears in the same category of census data as incorporated places. This distinguishes CDPs from other census classifications, such as minor civil divisions (MCDs), which are in a separate category.[5]

The population and demographics of the CDP are included in the data of county subdivisions containing the CDP. In no case is a CDP defined within the boundaries of what the Census Bureau regards to be an incorporated city, village or borough.[5] However, the Census Bureau considers towns in New England states and New York as well as townships in some other states as MCDs, even though they are incorporated municipalities in those states. Thus, CDPs may be defined within New England towns or spanning the boundaries of multiple towns.[5]

Purpose of designation

There are a number of reasons for the CDP designation:

- The area may be more urban than its surroundings, having a concentration of population with a definite residential nucleus, such as Whitmore Lake, Michigan, or Hershey, Pennsylvania.
- A formerly incorporated place may disincorporate or be partly annexed by a neighboring town, but the former town or a part of it may still be reported by the census as a CDP by meeting criteria for a CDP. An example is the former village of Covedale (village in Ohio), compared with Covedale (CDP), Ohio.
- The CDP designation may apply to large military bases (or parts of a military base) that are not within the boundaries of any existing community, such as Fort Campbell North and Fort Knox in Kentucky.[5]
- In some cases, a CDP may be defined for the urbanized area surrounding an incorporated municipality, but which is outside the municipal boundaries, for example Greater Galesburg, Michigan or Greater Upper Marlboro, Maryland.
- In other cases, the boundary of an incorporated place may bisect a recognized community. An example of this is Bostonia, California, which straddles the city limits of El Cajon. The USGS places the nucleus of Bostonia within El Cajon. The Bostonia CDP covers the greater El Cajon area in unincorporated San Diego County that is generally north of that part of Bostonia within El Cajon.
- The Census Bureau treats all townships as unincorporated places, even in those states where townships are incorporated under state law. This is so even in those states (i.e., Indiana, Michigan, Minnesota, New Jersey and South Dakota) where the Census Bureau acknowledges that "All townships are actively functioning governmental units."[9]
- In some states, a CDP may be defined within an incorporated municipality that (for the purposes of the census) is regarded as a minor civil division. For example, towns in Massachusetts and Connecticut are incorporated municipalities, but may also include both rural and urban areas. CDPs may be defined to describe urbanized areas within such municipalities, as in the case of North Amherst, Massachusetts.
- Some CDPs represent an aggregation of several nearby communities, for example Shorewood-Tower Hills-Harbert, Michigan or Egypt Lake-Leto, Florida. However, the Census Bureau will be discontinuing this method for the 2010 Census.[1]
- Hawaii is the only state that has no incorporated places recognized by the U.S. Census Bureau below the county level. All data for places in Hawaii reported by the census are CDPs.[5]

- In some states, the Census Bureau may designate an entire minor civil division (MCD) as a CDP (for example West Bloomfield Township, Michigan or Reading, Massachusetts). Such designations are used in states where the MCDs function with strong governmental authority and provide services equivalent to an incorporated municipality (New England, the Middle Atlantic States, Michigan, and Wisconsin). MCDs appear in a separate category in census data from places (i.e., incorporated places and CDPs); however, such MCDs strongly resemble incorporated places, and so CDPs coterminous with the MCDs are defined so that such places appear in both categories of census data. This practice will no longer be used in the 2010 Census.[1]

See also

- Census county division
- Designated place, a counterpart in the Canadian census
- ZIP Code Tabulation Area

Notes

[1] Census Designated Place (CDP) Program for the 2010 Census - Final Criteria (http://www.census.gov/geo/www/psap2010/cdp_criteria.html), *Federal Register*, February 13, 2008 (Volume 73, Number 30), accessed December 5, 2008.

[2] United States Census Area Description (http://www.census.gov/geo/www/cob/pl_metadata.html#cdp)

[3] "Cities with 100,000 or More Population in 2000 ranked by Population per Square Mile, 2000 in Alphabetic Order" (http://www.census.gov/statab/ccdb/cit1040a.txt). United States Census Bureau, Population Division. 2008-07-10. . Retrieved 2008-07-13.

[4] "Annual Estimates of the Population for Incorporated Places in Virginia" (http://www.census.gov/popest/cities/tables/SUB-EST2006-04-51.csv). US Census Bureau. . Retrieved 2008-09-11.

[5] Chapter 9 - Places (http://www.census.gov/geo/www/GARM/Ch9GARM.pdf). Retrieved 2010-09-30.

[6] U.S. Bureau of the Census, Census Designated Place (CDP) Program for the 2010 Census — Proposed Criteria (http://edocket.access.gpo.gov/2007/pdf/E7-6465.pdf), 72 *Federal Register* 17326-17329, April 6, 2007

[7] "Participant Statistical Areas Program (PSAP)" (http://www.census.gov/geo/www/psap2010/psap2010_main.html), U.S. Census Bureau, accessed 2008-03-09

[8] Cartographic Boundary Files (http://www.census.gov/geo/www/cob/pl_metadata.html#cdp) Cartogaphic Boundary Files. Retrieved 2010-09-30.

[9] http://www.census.gov/geo/www/GARM/Ch8GARM.pdf

References

- U.S. Census Bureau, Geography Division, " *Cartographic Boundary Files* (http://www.census.gov/geo/www/cob/pl_metadata.html#cdp)", Census-designated place. Cartographic Operations Branch, July 18, 2001.
- U.S. Census Bureau, " *Census 2000 Statistical Areas Boundary Criteria* (http://www.census.gov/geo/www/psapage.html#CDP)", Census Designated Places (CDPs) - Census 2000 Criteria.
- U.S. Census Bureau, Geographic Areas Reference Manual (http://www.census.gov/geo/www/garm.html), United States Department of Commerce.

Antioch Township, Lake County, Illinois

Antioch Township	
— Township —	
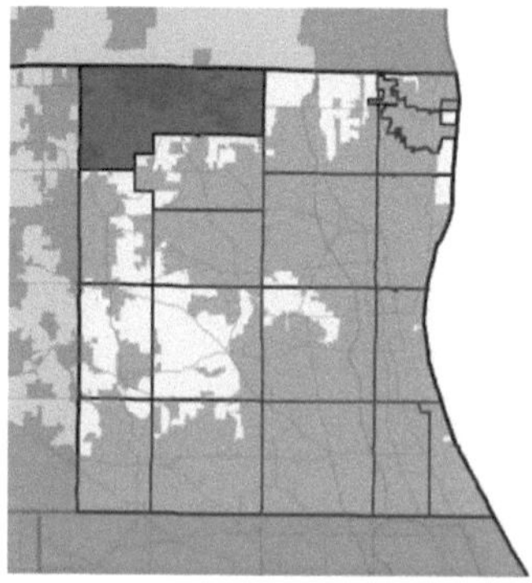	
Location in Lake County	
Coordinates: 42°27′52″N 88°06′41″W	
Country	United States
State	Illinois
County	Lake
Established	November 6, 1849
Government	
- Supervisor	
Area	
- Total	42.34 sq mi (109.65 km^2)
- Land	35.64 sq mi (92.3 km^2)
- Water	6.7 sq mi (17.35 km^2) 15.82%
Elevation	804 ft (245 m)
Population (2000)	
- Total	21879
- Density	613.9/sq mi (237/km^2)
Time zone	CST (UTC-6)
- Summer (DST)	CDT (UTC-5)
GNIS feature ID	0428594 [1]

Antioch Township is located in Lake County, Illinois. The population was 21,879 at the 2000 census.

Geography

Antioch Township covers an area of 42.34 square miles (109.7 km^2); of this, 6.7 square miles (17 km^2) or 15.82 percent is water. There are many lakes in this township, including Antioch Lake, Benet Lake, Bluff Lake, Channel Lake, Deer Lake, Dunns Lake, East Loon Lake, Grass Lake, Huntley Lake, Lake Catherine, Lake Marie, Loon Lake, Mud Lake, Petite Lake, Redwing Slough Lake, Silver Lake, Spring Lake and Turner Lake. [2]

Cities and towns

- Antioch
- Channel Lake
- Fox Lake (east quarter)
- Lake Catherine
- Lake Villa (north edge)

Adjacent townships

- Newport Township (east)
- Lake Villa Township (southeast)
- Grant Township (southwest)
- McHenry Township, McHenry County (southwest)
- Burton Township, McHenry County (west)

Cemeteries

The township contains three cemeteries: Grass Lake, Hickory Union and Hillside.

Major highways

- U.S. Route 45
- Illinois Route 59
- Illinois Route 83
- Illinois Route 173

Airports and landing strips

- Antioch Airport (historical)
- Donald Alfred Gade Airport
- Fox Lake Seaplane Base

Education

Antioch Community Consolidated School District 34

The schools in the district include:

Public elementary schools

- Antioch Elementary School, Antioch, grades 2-5;
- Emmons Grade School, Antioch, grades K-8;
- Grass Lake Elementary School, Antioch, grades PK-8;
- Hillcrest Elementary School, Antioch, grades PK-2;
- Oakland Elementary School, Antioch, grades 2-5;
- W C Petty Elementary School, Antioch, grades 2-5;

Public middle school

- Antioch Upper Grade School, Antioch, grades 6-8;

Public high school

- Antioch Community High School, Antioch, grades 9-12;

Private elementary/middle schools

- St Peter Catholic School, Antioch, grades KG-8;
- Faith Evangelical Lutheran School, Antioch, grades PK-8;

References

- U.S. Board on Geographic Names (GNIS) [3]
- United States Census Bureau cartographic boundary files [4]

[1] http://geonames.usgs.gov/pls/gnispublic/f?p=gnispq:3:::NO::P3_FID:0428594
[2] http://www.antioch.il.gov/
[3] http://geonames.usgs.gov/domestic/
[4] http://www.census.gov/geo/www/cob/

External links

- US Census (http://factfinder.census.gov/servlet/DatasetMainPageServlet?_lang=en&_)
- City-data.com (http://www.city-data.com/township/Antioch-Lake-IL.html)
- Illinois State Archives (http://www.cyberdriveillinois.com/departments/archives/irad/lake.html)

Lake County, Illinois

Lake County, Illinois	
 Location in the state of Illinois	
 Illinois's location in the U.S.	
Founded	1839
Seat	Waukegan
Area **- Total** **- Land** **- Water**	1367.95 sq mi (3543 km²) 447.56 sq mi (1159 km²) 920.39 sq mi (2384 km²), 67.28%
Population **- (2010)** **- Density**	703462 1572/sq mi (606.9/km²)
Website	www.lakecountyil.gov [1]

Lake County is a county in the northeastern corner of the state of Illinois, on the shore of Lake Michigan. In 2006, the United States Census Bureau estimated the population at 713,076. Its county seat is Waukegan[2] . The county is part of the Chicago metropolitan area. According to the 2000 Census, Lake County is the 31st richest county by per capita income. The lakefront communities of Lake Forest and Highland Park are part of the affluent North Shore area.

Originally part of McHenry County, the townships comprising Lake County were carved out into a separate county in 1839.

Naval Station Great Lakes is located in the city of North Chicago in Lake County. It is the United States Navy's Headquarters Command for training, and the Navy's only recruit training center.

History

The county, which was primarily unsettled prairie and was still home to its native Potawatomi Indians, was created by the Illinois State Legislature in 1839. At that time, Libertyville, then known as Independence Grove, was the first county seat. In 1841, however, the county's residents voted to move the county government to Little Fort, now Waukegan, where the commissioners had purchased a section of land from the state. Lake County's first courthouse was built on part of that land in 1844 and the remainder was sold to pay for the $4,000 construction cost.

The Caspar Ott Cabin, built in 1837, is the oldest structure in Lake County.

The county's first courthouse was used solely for court sessions and the jail, but in 1853, commissioners constructed a building to accommodate county administration offices and house records. When fire damaged the courthouse on October 19, 1875, the county records were saved because they were in the adjacent building.

Lake County at the time of its creation in 1839

After the fire, proposals were made to move the county seat to Highland Park, Libertyville or another site in central Lake County. The county commissioners, however, decided to rebuild in Waukegan. The east half of the building was reconstructed at a cost of $45,000. In 1895, the first jail building was added to the government complex and a west addition was added to the courthouse in 1922. By 1938, county commissioners saw a need for additional space and approved the addition of a 5th Floor. This courthouse, however, was demolished in 1967 to make room for a new high-rise administration building, which was completed with the addition of the jail in 1969 and courts in 1970.

Shortly thereafter, the Lake County Board commissioned the construction of a multi-faceted justice facility and ground was broken in 1986 for the Robert H. Babcox Justice Center, named in memory of Sheriff Babcox, who served as Lake County Sheriff from 1982-1988. The justice center, which houses the county jail, work release program, sheriff's administration offices and three courtrooms, was finished in 1989 at a cost of $29.6 million.

Additional county government facilities have been built or expanded throughout Lake County, including the Coroner's Office, Health Department/Community Health Center facilities, Division of Transportation, Public Works and Winchester House. Lake County government services extend throughout the county's 470 square miles (1200 km^2).

The historic Half Day Inn, a tavern/restaurant, was constructed in 1843. This structure, once located at the corner of Milwaukee Avenue and Rte. 45/ Olde Half Day Road, had long been the oldest structure in Lake County (older than the Casper Ott Cabin) until it was demolished in 2007 (in the eyes of some residents unnecessarily) to make way for retail space, condominiums, and a retention pond.

Demographics

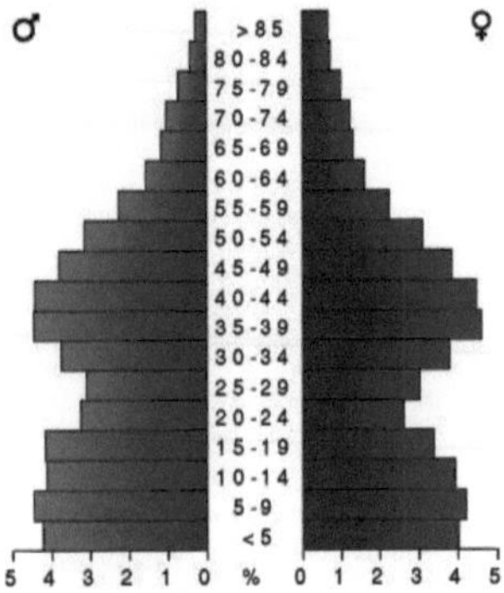

2000 census age pyramid for Lake County.

Historical populations		
Census	Pop.	%±
1840	2634	—
1850	14226	440.1%
1860	18257	28.3%
1870	21014	15.1%
1880	21296	1.3%
1890	24235	13.8%
1900	34504	42.4%
1910	55058	59.6%
1920	74285	34.9%
1930	104387	40.5%
1940	121094	16.0%
1950	179097	47.9%
1960	293656	64.0%
1970	382638	30.3%
1980	440372	15.1%
1990	516418	17.3%
2000	644356	24.8%
2010	703462	9.2%

As of the census[3] of 2000, there were 644,356 people, 216,297 households, and 163,953 families residing in the county. The population density was 1,440 people per square mile (556/km²). There were 225,919 housing units at an average density of 505 per square mile (195/km²). The racial makeup of the county was 80.11% White, 6.94% Black or African American, 0.28% Native American, 3.90% Asian, 0.05% Pacific Islander, 6.72% from other races, and 2.01% from two or more races. 14.39% of the population were Hispanic or Latino of any race. 17.0% were of

German, 9.0% Irish, 7.2% Polish, 6.4% Italian and 5.0% American ancestry according to Census 2000. 78.7% spoke English, 12.9% Spanish and 1.0% Russian as their first language.

The Census estimates of 2005 showed Lake County with 68.6% of its residents being non-Hispanic whites, 6.8% African American and 5.5% Asian. 18.2% were Hispanic or Latino of any race.[4]

According to the 2000 census there were 216,297 households out of which 42.10% had children under the age of 18 living with them, 63.20% were married couples living together, 9.20% had a female householder with no husband present, and 24.20% were non-families. 19.70% of all households were made up of individuals and 6.20% had someone living alone who was 65 years of age or older. The average household size was 2.88 and the average family size was 3.33.

In the county the population was spread out with 29.40% under the age of 18, 8.90% from 18 to 24, 31.60% from 25 to 44, 21.60% from 45 to 64, and 8.50% who were 65 years of age or older. The median age was 34 years. For every 100 females there were 101.20 males. For every 100 females age 18 and over, there were 99.30 males. Waukegan and North Chicago account for 20% of the county's population.

The median income for a household in the county was $66,973, and the median income for a family was $76,424 (these figures had risen to $76,940 and $88,851 respectively as of a 2007 estimate[5]). Males had a median income of $50,789 versus $33,458 for females. The per capita income for the county was $32,102. About 4.00% of families and 5.70% of the population were below the poverty line, including 7.00% of those under age 18 and 4.80% of those age 65 or over. The 2000 Census reported that Lake County had the second median highest household income of any county in the state, behind DuPage County.

Geography

According to the 2000 census, the county has a total area of 1367.95 square miles (3543.0 km^2), of which 447.56 square miles (1159.2 km^2) (or 32.72%) is land and 920.39 square miles (2383.8 km^2) (or 67.28%) is water,[6] much of it in Lake Michigan

A satellite view of the Lake County district

State parks

- Illinois Beach State Park
- North Point Marina
- Volo Bog State Natural Area
- Chain O'Lakes State Park

Lakes

Besides Lake Michigan, lakes in the county include:

- Lake Amy
- Antioch Lake
- Ashley Lake
- Bangs Lake
- Lake Barrington
- Benet Lake
- Big Bear Lake
- Bluff Lake
- Brandenburg Lake
- Bresen Lake
- Butler Lake
- Lake Catherine
- Cedar Lake
- Channel Lake
- Lake Charles
- Countryside Lake
- Cranberry Lake
- Crooked Lake
- Cross Lake
- Davis Lake
- Davlins Pond
- Dead Lake
- Deep Lake
- Deer Lake
- Diamond Lake
- Druce Lake
- Duck Lake
- Dunns Lake
- East Loon Lake
- Echo Lake
- Lake Eleanor
- Elmwood Farms Lake
- Lake Fairfield
- Fish Lake
- Forest Lake
- Fourth Lake
- Fox Lake
- Gages Lake
- Lake Germaine
- Grandwood Lake
- Grass Lake
- Grassy Lake
- Grayslake
- Harvey Lake
- Hastings Lake
- Hendrick Lake
- Highland Lake
- Homer White Lake
- Honey Lake
- Huntley Lake
- Island Lake
- Lambs Lake
- Leo Lake
- Liberty Lake
- Lake Linden
- Little Bear Lake
- Loch Lomond
- Long Lake
- Loon Lake
- Lake Louise
- Lucky Lake
- Lake Marie
- Miltmore Lake
- Minear Lake
- Mud Lake
- Lake Naomi
- Nippersink Lake
- North Tower Lake
- Petite Lake
- Pistakee Lake
- Lake Potomac
- Redhead Lake
- Redwing Slough Lake
- Round Lake
- Saint Marys Lake
- Sand Lake
- Sand Pond
- Schreiber Lake
- Shady Lane Resort Lake
- Lake Sheree
- Silver Lake
- Slocum Lake
- Slough Lake
- Spring Lake
- Sterling Lake
- Sullivan Lake
- Sun Lake
- Sylvan Lake
- Third Lake
- Timber Lake
- Tower Lake
- Turner Lake
- Twin Lakes
- Valley Lake
- Waterford Lake
- West Lake
- White Lake
- Wooster Lake
- Lake Zurich

Forest preserves and natural areas

Lake County's forest preserves and natural areas are administered by the Lake County Forest Preserves district. These facilities include traditional nature preserves, such as the Ryerson Conservation Area, as well as golf courses and historic homes, such as the Adlai Stevenson historic home.[7] A long north-south string of the preserves in Lake County, including Half Day Woods, Old School Forest Preserve, Independence Grove, and Van Patten Woods, form the Des Plaines River Greenway, which contains the Des Plaines River Trail, a popular place for walking, running, and biking.

Adjacent counties

- Kenosha County, Wisconsin - north
- Allegan County, Michigan - northeast[1]
- Van Buren County, Michigan - east[1]
- Berrien County, Michigan - southeast[1]
- Cook County, Illinois - south
- McHenry County, Illinois - west

[1] Boundary lies entirely in Lake Michigan.

Major highways

- Interstate 94
- U.S. Highway 12
- U.S. Highway 14
- U.S. Highway 41
- U.S. Highway 45
- Illinois Route 21
- Illinois Route 22
- Illinois Route 43
- Illinois Route 53
- Illinois Route 59
- Illinois Route 60
- Illinois Route 83
- Illinois Route 120
- Illinois Route 131
- Illinois Route 132
- Illinois Route 137
- Illinois Route 134
- Illinois Route 173

Townships

The county is divided into eighteen townships.

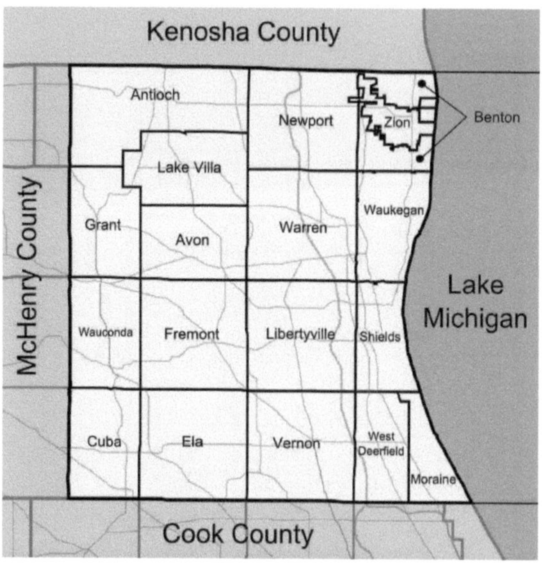

Lake County townships (clickable)

Township	Population (2000)
Waukegan	92,805
Vernon	65,355
Warren	59,424
Avon	54,957
Libertyville	48,904
Shields	43,382
Ela	39,688
Moraine	34,538
Lake Villa	33,721
West Deerfield	31,794
Fremont	23,955
Zion	22,866
Antioch	21,879
Benton	17,413
Grant	17,397
Wauconda	16,387
Cuba	15,749
Newport	4,142

Cities, villages and census-designated places

Cities

Name	Incorporation year	Population (2000)	Additional information
Waukegan	1829	87,901	-
North Chicago	1895	35,918	-
Highland Park	1869	31,365	-
Zion	1901	22,866	-
Lake Forest	1861	20,059	-
Park City	1958	6,637	-
Highwood	1887	4,143	-

Villages

Name	Incorporation year	Population (2000)	Additional information
Buffalo Grove	1958	45,000	Partly in Cook County.
Mundelein	1909	30,935	-
Gurnee	1928	28,834	-
Round Lake Beach	1937	25,859	-
Libertyville	1882	21,760	-
Vernon Hills	1958	20,120	-
Grayslake	1895	18,506	-
Deerfield	1903	18,420	Partly in Cook County.
Lake Zurich	1896	18,104	-
Lindenhurst	1956	12,539	-
Barrington	1865	10,168	Partly in Cook County
Beach Park	1988	10,072	-
Wauconda	1877	9,448	-
Fox Lake	1907	9,178	Partly in McHenry County.
Antioch	1892	8,788	-
Island Lake	1952	8,153	Partly in McHenry County.
Long Grove	1956	6,735	-
Winthrop Harbor	1901	6,670	-
Lincolnshire	1957	6,108	-
Lake Bluff	1895	6,056	-
Round Lake Park	1947	6,038	-
Hawthorn Woods	1958	6,002	-
Lake Villa	1901	5,864	-
Round Lake	1909	5,842	-
Fox River Grove	1919	4,862	Primarily in McHenry County.
Lake Barrington	1959	4,757	-
Barrington Hills	1957	3,915	Partly in Cook, Kane and McHenry County.
Riverwoods	1959	3,843	-
Green Oaks	1960	3,572	-
Kildeer	1958	3,460	-
Deer Park	1957	3,102	Partly in Cook County.
Wadsworth	1962	3,083	-
North Barrington	1959	2,918	-
Lakemoor	1938	2,788	Partly in McHenry County.
Hainesville	1846	2,129	-
Bannockburn	1924	1,429	-
Third Lake	1959	1,355	-
Round Lake Heights	1960	1,347	-

Tower Lakes	1966	1,310	-
Port Barrington	1969	788	Partly in McHenry County.
Mettawa	1960	367	-
Old Mill Creek	1959	251	-
Indian Creek	1958	194	-
Volo	1993	180	-

Unincorporated communities

Name	Population (2000)	Additional information
Gages Lake	10,415	-
Ingleside	9,286	-
Grandwood Park	4,521	-
Long Lake	3,356	-
Venetian Village	3,082	-
Prairie View	3,079	-
Fox Lake Hills	2,561	-
Channel Lake	1,785	-
Forest Lake	1,530	-
Lake Catherine	1,490	-
Ivanhoe	?	-
Knollwood	?	-
Rondout	?	-
Russell	?	-
Wildwood	?	-

Sports

The following sports teams play in Lake County:

- Lake County Fielders baseball

Sites of interest

Amusement Parks

- Six Flags Great America located in Gurnee, Illinois.

Six Flags Great America

Harbors

- North Point Marina located in Winthrop Harbor, Illinois.
- Waukegan Harbor located in Waukegan, Illinois.

Museums

- Volo Auto Museum - an automobile museum in Volo, Illinois. It is part of a larger antique mall.

Shopping

- Gurnee Mills Mall located in Gurnee, Illinois.
- Westfield Hawthorn Center located in Vernon Hills, Illinois.

Other

- Naval Station Great Lakes
- Ravinia Park located in Highland Park, Illinois.
- Tempel Farms - Lipizzan Horse Stables and Outdoor Arena

Notable natives and residents

Notable people who are natives of or have lived in Lake County include:

- Jack Benny (February 14, 1894 - December 26, 1974) - American entertainer. Comedian, actor and musician, Benny was one of America's greatest stars of radio and television, and also appeared in many films. Benny was raised in Waukegan, Illinois.
- Ray Bradbury (born August 22, 1920) - American mainstream, fantasy, horror, science fiction, and mystery writer. Best known for his dystopian novel *Fahrenheit 451* and *The Martian Chronicles*, Bradbury is widely considered one of the greatest and most popular American writers of speculative fiction of the twentieth century. Bradbury was born in Waukegan, Illinois.
- Barry Bradford, teacher, author, public speaker, famous for helping reopen the Mississippi Burning Case, and the Clyde Kennard Case. National Teacher Of The Year [8] , Illinois Teacher Of The Year, and winner of the Golden Apple Award for Excellence In Teaching[9] , as well as a Presidential Citation For Civillian Service. [10] is a long time resident of Deerfield, Illinois
- Gary Coleman (February 8, 1968 - May 28, 2010) - American actor. Best known for his role as Arnold Jackson in the American sitcom *Diff'rent Strokes* (1978–1986). Coleman was born in Zion, Illinois.

- Marlon Brando (April 3, 1924 – July 1, 2004) American actor. As a young sex symbol, he is best known for his roles in *A Streetcar Named Desire*, *On the Waterfront* and *Guys and Dolls*. Brando and his family moved to Libertyville, Illinois where he lived from 1937 until 1942.
- Michael Jordan (born February 17, 1963) - a retired American professional basketball player and active businessman. Jordan is widely considered one of the greatest and most popular basketball players of all time. As of 2010, Jordan lives in Highland Park, Illinois.
- Vince Vaughn (born March 28, 1970) - American actor. Best known for his roles in *Swingers* and *Wedding Crashers*. Vaughn grew up in Buffalo Grove, Illinois. He then moved to Lake Forest, Illinois, where he graduated from Lake Forest High School in 1988.

Lake County in popular culture

The most notable movies which filmed through the years at Lake county are:

- *Ordinary People* (1980) - the movie was filmed in Highwood, Highland Park, Lake Bluff, Lake Forest, Northbrook and Wilmette.
- *Class* (1983) - the movie was filmed at Lake Forest College in Lake Forest and other locations in Chicago.
- *Risky Business* (1983) - the movie was filmed in Deerfield, Highland Park and Skokie.
- *Chain Reaction* (1996) - the movie has scenes filmed at a famous Lake Bluff estate and was largely shot in downtown Chicago.

See also

- List of school districts in Lake County, Illinois
- National Register of Historic Places listings in Lake County, Illinois

References

[1] http://www.lakecountyil.gov
[2] "Find a County" (http://www.naco.org/Counties/Pages/FindACounty.aspx). National Association of Counties. . Retrieved 2011-06-07.
[3] "American FactFinder" (http://factfinder.census.gov). United States Census Bureau. . Retrieved 2008-01-31.
[4] Lake County QuickFacts from the US Census Bureau (http://quickfacts.census.gov/qfd/states/17/17097.html)
[5] http://factfinder.census.gov/servlet/ACSSAFFFacts?_event=ChangeGeoContext&geo_id=05000US17097&_geoContext=01000US%7C04000US17%7C05000US17093&_street=&_county=lake&_cityTown=lake&_state=04000US17&_zip=&_lang=en&_sse=on&ActiveGeoDiv=geoSelect&_useEV=&pctxt=fph&pgsl=010&_submenuId=factsheet_1&ds_name=ACS_2007_3YR_SAFF&_ci_nbr=null&qr_name=null&reg=null%3Anull&_keyword=&_industry=
[6] "Census 2000 U.S. Gazetteer Files: Counties" (http://www.census.gov/tiger/tms/gazetteer/county2k.txt). United States Census. . Retrieved 2011-02-13.
[7] "Lake County Forest Preserves" (http://www.lcfpd.org/index.cfm). .
[8] www.oah.org/awards/awards.tachau.winners.html
[9] www.goldenapple.org/pages/academy_directory/26.php
[10] http://www.barrybradford.com/2.html.

- Forstall, Richard L. (editor) (1996). *Population of states and counties of the United States: 1790 to 1990 : from the twenty-one decennial censuses*. United States Department of Commerce, Bureau of the Census, Population Division. ISBN 0-934213-48-8.

External links

- Lake County, IL official web site (http://www.lakecountyil.gov/)
- Lake County Forest Preserve District official web site (http://www.LCFPD.org/)
- Preservation Foundation of the Lake County Forest Preserves (http://www.LCFPD.org/donate)
- Lake County Animal Control official web site (http://www.co.lake.il.us/health/ehs/animal.asp)
- Lake County Online web site (http://www.lakeonline.com/)
- Pioneer Press Lake County Blog (http://blogs.pioneerlocal.com/lake/)
- Lake County, Illinois, CVB - Official Travel Site (http://lakecounty.org/)
- About Lake County (http://www.lakecountyil.gov/ABOUTOURCOUNTY/HISTORY/default.htm/)

United_States

United States of America

Motto: In God We Trust (official)
E Pluribus Unum (traditional)
(Latin: Out of Many, One)

Anthem: "The Star-Spangled Banner"

Capital		Washington, D.C. 38°53′N 77°01′W
Largest city		New York City
Official language(s)		None at federal level[a]
National language		English (*de facto*)[b]
Demonym		American
Government		Federal presidential constitutional republic
-	President	Barack Obama (D)

-	Vice President	Joe Biden (D)
-	Speaker of the House	John Boehner (R)
-	Chief Justice	John Roberts
Legislature		Congress
-	Upper House	Senate
-	Lower House	House of Representatives
Independence from the Kingdom of Great Britain		
-	Declared	July 4, 1776
-	Recognized	September 3, 1783
-	Current constitution	June 21, 1788
Area		
-	Total	9826675 km^2 [1] [c](3rd/4th) 3794101 sq mi
-	Water (%)	6.76
Population		
-	2011 estimate	312421000[2] (3rd)
-	Density	33.7/km^2 87.4/sq mi
GDP (PPP)		2011 estimate
-	Total	$15.065 trillion[3] (1st)
-	Per capita	$48,147[3] (8th)
GDP (nominal)		2011 estimate
-	Total	$15.065 trillion[4] (1st)
-	Per capita	$48,147[3] (15th)
Gini (2007)		45.0[1] (39th)
HDI (2010)		▲ 0.902[5] (very high) (4th)
Currency		United States dollar ($) (USD)
Time zone		(UTC−5 to −10)
-	Summer (DST)	(UTC−4 to −10)
Date formats		m/d/yy (AD)
Drives on the		right
Internet TLD		.us .gov .mil .edu
Calling code		+1

a. English is the official language of at least 28 states—some sources give a higher figure, based on differing definitions of "official".[6] English and Hawaiian are both official languages in the state of Hawaii.

b. English is the *de facto* language of American government and the sole language spoken at home by 80% of Americans age five and older. Spanish is the second most commonly spoken language.

c. Whether the United States or the People's Republic of China is larger is disputed. The figure given is from the U.S. Central Intelligence Agency's *World Factbook*. Other sources give smaller figures. All authoritative calculations of the country's size include only the 50 states and the District of Columbia, not the territories.

d. The population estimate includes people whose usual residence is in the fifty states and the District of Columbia, including noncitizens. It does not include either those living in the territories, amounting to more than 4 million U.S. citizens (most in Puerto Rico), or U.S. citizens living outside the United States.

The **United States of America** (also called the **United States**, the **States**, the **U.S.**, the **USA**, and **America**) is a federal constitutional republic comprising fifty states and a federal district. The country is situated mostly in central North America, where its forty-eight contiguous states and Washington, D.C., the capital district, lie between the Pacific and Atlantic Oceans, bordered by Canada to the north and Mexico to the south. The state of Alaska is in the northwest of the continent, with Canada to the east and Russia to the west across the Bering Strait. The state of Hawaii is an archipelago in the mid-Pacific. The country also possesses several territories in the Pacific and Caribbean.

At 3.79 million square miles (9.83 million km^2) and with over 312 million people, the United States is the third or fourth largest country by total area, and the third largest by both land area and population. It is one of the world's most ethnically diverse and multicultural nations, the product of large-scale immigration from many countries.[7] The U.S. economy is the world's largest national economy, with an estimated 2010 GDP of $14.53 trillion (23% of nominal global GDP and over 19% of global GDP at purchasing-power parity).[4] [8]

Indigenous peoples descended from forebears who migrated from Asia have inhabited what is now the mainland United States for many thousands of years. This Native American population was greatly reduced by disease and warfare after European contact. The United States was founded by thirteen British colonies located along the Atlantic seaboard. On July 4, 1776, they issued the Declaration of Independence, which proclaimed their right to self-determination and their establishment of a cooperative union. The rebellious states defeated the British Empire in the American Revolution, the first successful colonial war of independence.[9] The current United States Constitution was adopted on September 17, 1787; its ratification the following year made the states part of a single republic with a strong central government. The Bill of Rights, comprising ten constitutional amendments guaranteeing many fundamental civil rights and freedoms, was ratified in 1791.

Through the 19th century, the United States displaced native tribes, acquired the Louisiana territory from France, Florida from Spain, part of the Oregon Country from the United Kingdom, Alta California and New Mexico from Mexico, and Alaska from Russia, and annexed the Republic of Texas and the Republic of Hawaii. Disputes between the agrarian South and industrial North over the expansion of the institution of slavery and states' rights provoked the Civil War of the 1860s. The North's victory prevented a permanent split of the country and led to the end of legal slavery in the United States. By the 1870s, its national economy was the world's largest.[10] The Spanish–American War and World War I confirmed the country's status as a military power. It emerged from World War II as the first country with nuclear weapons and a permanent member of the United Nations Security Council. The end of the Cold War and the dissolution of the Soviet Union left the United States as the sole superpower. The country accounts for 41% of global military spending,[11] and it is a leading economic, political, and cultural force in the world.[12]

Etymology

In 1507, German cartographer Martin Waldseemüller produced a world map on which he named the lands of the Western Hemisphere "America" after Italian explorer and cartographer Amerigo Vespucci.[13] The former British colonies first used the country's modern name in the 1776 Declaration of Independence, the "unanimous Declaration of the thirteen united States of America".[14] On November 15, 1777, the Second Continental Congress adopted the Articles of Confederation, which states, "The Stile of this Confederacy shall be 'The United States of America.'" The Franco-American treaties of 1778 used "United States of North America", but from July 11, 1778, "United States of America" was used on the country's bills of exchange, and it has been the official name ever since.[15]

The short form "United States" is also standard. Other common forms include the "U.S.", the "USA", and "America". Colloquial names include the "U.S. of A." and, internationally, the "States". "Columbia", a once popular name for the United States, derives from Christopher Columbus; it appears in the name "District of Columbia".

The standard way to refer to a citizen of the United States is as an "American". Although "United States" is the official appositional term, "American" and "U.S." are more commonly used to refer to the country adjectivally ("American values", "U.S. forces"). "American" is rarely used in English to refer to people not connected to the United States.[16]

The phrase "United States" was originally treated as plural—e.g., "the United States are"—including in the Thirteenth Amendment to the United States Constitution, ratified in 1865. It became common to treat it as singular—e.g., "the United States is"—after the end of the Civil War. The singular form is now standard; the plural form is retained in the idiom "these United States".[17]

Geography and environment

The land area of the contiguous United States is approximately 1900 million acres (km^2). Alaska, separated from the contiguous United States by Canada, is the largest state at 365 million acres (km^2). Hawaii, occupying an archipelago in the central Pacific, southwest of North America, has just over 4 million acres (16000 km^2).[18] The United States is the world's third or fourth largest nation by total area (land and water), ranking behind Russia and Canada and just above or below China. The ranking varies depending on how two territories disputed by China and India are counted and how the total size of the United States is measured: calculations range from 3676486 square miles (km^2)[19] to 3717813 square miles (km^2)[20] to 3794101 square miles (km^2).[1] Including only land area, the United States is third in size behind Russia and China, just ahead of Canada.[21]

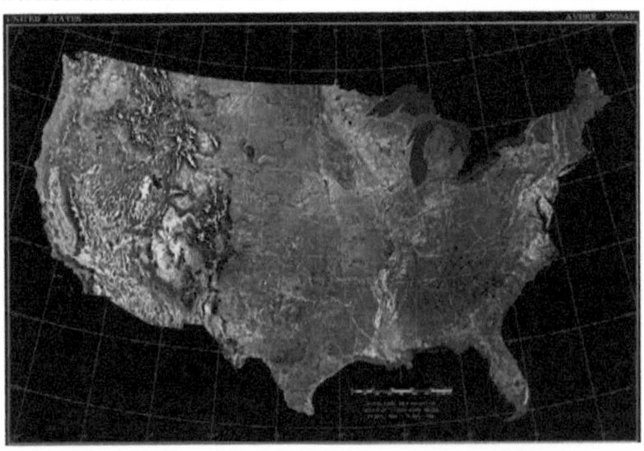

Satellite image showing topography of the contiguous United States

The coastal plain of the Atlantic seaboard gives way further inland to deciduous forests and the rolling hills of the Piedmont. The Appalachian Mountains divide the eastern seaboard from the Great Lakes and the grasslands of the Midwest. The Mississippi–Missouri River, the world's fourth longest river system, runs mainly north–south through the heart of the country. The flat, fertile prairie of the Great Plains stretches to the west, interrupted by a highland region in the southeast. The Rocky Mountains, at the western edge of the Great Plains, extend north to south across the country, reaching altitudes higher than 14,000 feet (4,300 m) in Colorado. Farther west are the rocky Great Basin and deserts such as the Mojave. The Sierra Nevada and Cascade mountain ranges run close to the Pacific coast. At 20,320 feet (6,194 m), Alaska's Mount McKinley is the tallest peak in the country and in North America. Active volcanoes are common throughout Alaska's Alexander and Aleutian Islands, and Hawaii consists of volcanic islands. The supervolcano underlying Yellowstone National Park in the Rockies is the continent's largest volcanic feature.[22]

The United States, with its large size and geographic variety, includes most climate types. To the east of the 100th meridian, the climate ranges from humid continental in the north to humid subtropical in the south. The southern tip of Florida is tropical, as is Hawaii. The Great Plains west of the 100th meridian are semi-arid. Much of the Western mountains are alpine. The climate is arid in the Great Basin, desert in the Southwest, Mediterranean in coastal California, and oceanic in coastal Oregon and Washington and southern Alaska. Most of Alaska is subarctic or polar. Extreme weather is not uncommon—the states bordering the Gulf of Mexico are prone to hurricanes, and most of the world's tornadoes occur within the country, mainly in the Midwest's Tornado Alley.[23]

The bald eagle, national bird of the United States since 1782

The U.S. ecology is considered "megadiverse": about 17,000 species of vascular plants occur in the contiguous United States and Alaska, and over 1,800 species of flowering plants are found in Hawaii, few of which occur on the mainland.[24] The United States is home to more than 400 mammal, 750 bird, and 500 reptile and amphibian species.[25] About 91,000 insect species have been described.[26] The Endangered Species Act of 1973 protects threatened and endangered species and their habitats, which are monitored by the United States Fish and Wildlife Service. There are fifty-eight national parks and hundreds of other federally managed parks, forests, and wilderness areas.[27] Altogether, the government owns 28.8% of the country's land area.[28] Most of this is protected, though some is leased for oil and gas drilling, mining, logging, or cattle ranching; 2.4% is used for military purposes.[28]

Political divisions

Further information: Territorial evolution of the United States and United States territorial acquisitions

The United States is a federal union of fifty states. The original thirteen states were the successors of the thirteen colonies that rebelled against British rule. Early in the country's history, three new states were organized on territory separated from the claims of the existing states: Kentucky from Virginia; Tennessee from North Carolina; and Maine from Massachusetts. Most of the other states have been carved from territories obtained through war or purchase by the U.S. government. One set of exceptions comprises Vermont, Texas, and Hawaii: each was an independent republic before joining the union. During the American Civil War, West Virginia broke away from Virginia. The most recent state—Hawaii—achieved statehood on August 21, 1959. The states do not have the right to secede from the union.

The states compose the vast bulk of the U.S. land mass; the two other areas considered integral parts of the country are the District of Columbia, the federal district where the capital, Washington, is located; and Palmyra Atoll, an uninhabited but incorporated territory in the Pacific Ocean. The United States also possesses five major overseas territories: Puerto Rico and the United States Virgin Islands in the Caribbean; and American Samoa, Guam, and the Northern Mariana Islands in the Pacific.[29] Those born in the major territories (except for American Samoa) possess U.S. citizenship.[30] American citizens residing in the territories have many of the same rights and responsibilities as citizens residing in the states; however, they are generally exempt from federal income tax, may not vote for president, and have only nonvoting representation in the U.S. Congress.[31]

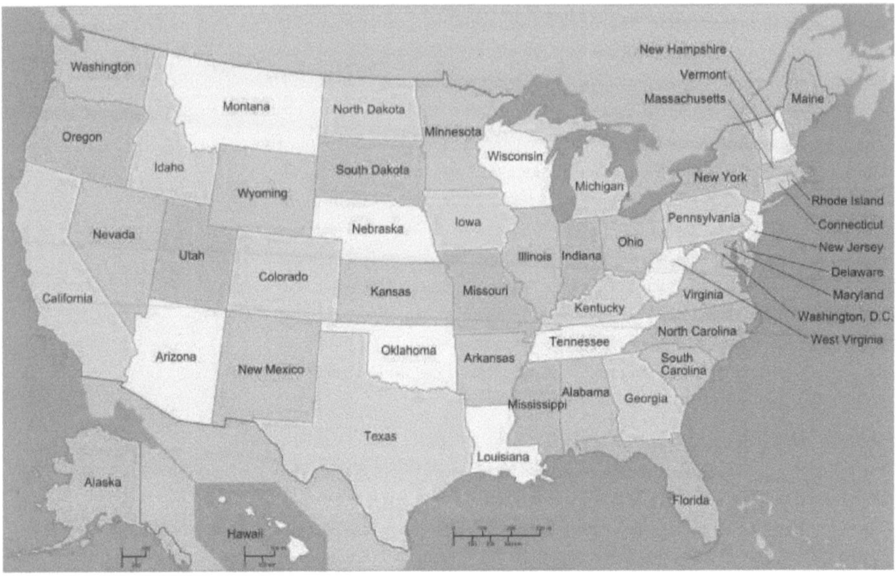

History

Native American and European settlement

The indigenous peoples of the U.S. mainland, including Alaska Natives, are believed to have migrated from Asia, beginning between 12,000 and 40,000 years ago.[32] Some, such as the pre-Columbian Mississippian culture, developed advanced agriculture, grand architecture, and state-level societies. After Europeans began settling the Americas, many millions of indigenous Americans died from epidemics of imported diseases such as smallpox.[33]

The *Mayflower* transported Pilgrims to the New World in 1620, as depicted in William Halsall's *The Mayflower in Plymouth Harbor*, 1882

In 1492, Genoese explorer Christopher Columbus, under contract to the Spanish crown, reached several Caribbean islands, making first contact with the indigenous people. On April 2, 1513, Spanish conquistador Juan Ponce de León landed on what he called "La Florida"—the first documented European arrival on what would become the U.S. mainland. Spanish settlements in the region were followed by ones in the present-day southwestern United States that drew thousands through Mexico. French fur traders established outposts of New France around the Great Lakes; France eventually claimed much of the North American interior, down to the Gulf of Mexico. The first successful English settlements were the Virginia Colony in Jamestown in 1607 and the Pilgrims' Plymouth Colony in 1620. The 1628 chartering of the Massachusetts Bay Colony resulted in a wave of migration; by 1634, New England had been settled by some 10,000 Puritans. Between the late 1610s and the American Revolution, about 50,000 convicts were shipped to Britain's American colonies.[34] Beginning in 1614, the Dutch settled along the lower Hudson River, including New Amsterdam on Manhattan Island.

In 1674, the Dutch ceded their American territory to England; the province of New Netherland was renamed New York. Many new immigrants, especially to the South, were indentured servants—some two-thirds of all Virginia immigrants between 1630 and 1680.[35] By the turn of the 18th century, African slaves were becoming the primary source of bonded labor. With the 1729 division of the Carolinas and the 1732 colonization of Georgia, the thirteen British colonies that would become the United States of America were established. All had local governments with elections open to most free men, with a growing devotion to the ancient rights of Englishmen and a sense of self-government stimulating support for republicanism. All legalized the African slave trade. With high birth rates, low death rates, and steady immigration, the colonial population grew rapidly. The Christian revivalist movement of the 1730s and 1740s known as the Great Awakening fueled interest in both religion and religious liberty. In the French and Indian War, British forces seized Canada from the French, but the francophone population remained politically isolated from the southern colonies. Excluding the Native Americans (popularly known as "American Indians"), who were being displaced, those thirteen colonies had a population of 2.6 million in 1770, about one-third that of Britain; nearly one in five Americans were black slaves.[36] Though subject to British taxation, the American colonials had no representation in the Parliament of Great Britain.

Independence and expansion

Declaration of Independence, by John Trumbull, 1817–18

Tensions between American colonials and the British during the revolutionary period of the 1760s and early 1770s led to the American Revolutionary War, fought from 1775 to 1781. On June 14, 1775, the Continental Congress, convening in Philadelphia, established a Continental Army under the command of George Washington. Proclaiming that "all men are created equal" and endowed with "certain unalienable Rights", the Congress adopted the Declaration of Independence, drafted largely by Thomas Jefferson, on July 4, 1776. That date is now celebrated annually as America's Independence Day. In 1777, the Articles of Confederation established a weak confederal government that operated until 1789.

After the British defeat by American forces assisted by the French and Spanish, Great Britain recognized the independence of the United States and the states' sovereignty over American territory west to the Mississippi River. Those wishing to establish a strong national government with powers of taxation organized a constitutional convention in 1787. The United States Constitution was ratified in 1788, and the new republic's first Senate, House of Representatives, and president—George Washington—took office in 1789. The Bill of Rights, forbidding federal restriction of personal freedoms and guaranteeing a range of legal protections, was adopted in 1791.

Attitudes toward slavery were shifting; a clause in the Constitution protected the transatlantic slave trade only until 1808. The Northern states abolished slavery between 1780 and 1804, leaving the slave states of the South as defenders of the "peculiar institution". The Second Great Awakening, beginning about 1800, made evangelicalism a force behind various social reform movements, including abolitionism.

Territorial acquisitions by date

Americans' eagerness to expand westward prompted a long series of Indian Wars. The Louisiana Purchase of French-claimed territory under President Thomas Jefferson in 1803 almost doubled the nation's size.[37] The War of 1812, declared against Britain over various grievances and fought to a draw, strengthened U.S. nationalism. A series of U.S. military incursions into Florida led Spain to cede it and other Gulf Coast territory in 1819. The Trail of Tears in the 1830s exemplified the Indian removal policy that stripped the native peoples of their land. The United States annexed the Republic of Texas in 1845, amid a period when the concept of Manifest Destiny was becoming popular.[38] The 1846 Oregon Treaty with Britain led to U.S. control of the present-day American Northwest. The U.S. victory in the Mexican-American War resulted in the 1848 cession of California and much of the present-day American Southwest. The California Gold Rush of 1848–49 further spurred western migration. New railways made relocation easier for settlers and increased conflicts with Native Americans. Over a half-century, up to 40 million American bison, or buffalo, were slaughtered for skins and meat and to ease the railways' spread. The loss of the buffalo, a primary resource for the plains Indians, was an existential blow to many native cultures.

Civil War and industrialization

Battle of Gettysburg, lithograph by Currier & Ives, ca. 1863

Tensions between slave and free states mounted with arguments over the relationship between the state and federal governments, as well as violent conflicts over the spread of slavery into new states. Abraham Lincoln, candidate of the largely antislavery Republican Party, was elected president in 1860. Before he took office, seven slave states declared their secession—which the federal government maintained was illegal—and formed the Confederate States of America. With the Confederate attack upon Fort Sumter, the Civil War began and four more slave states joined the Confederacy. Lincoln's Emancipation Proclamation in 1863 declared slaves in the Confederacy to be free. Following the Union victory in 1865, three amendments to the U.S. Constitution ensured freedom for the nearly four million African Americans who had been slaves,[39] made them citizens, and gave them voting rights. The war and its resolution led to a substantial increase in federal power.[40] The war remains the deadliest conflict in American history, resulting in the deaths of 620,000 soldiers.[41]

Immigrants at Ellis Island, New York Harbor, 1902

After the war, the assassination of Lincoln radicalized Republican Reconstruction policies aimed at reintegrating and rebuilding the Southern states while ensuring the rights of the newly freed slaves. The resolution of the disputed 1876 presidential election by the Compromise of 1877 ended Reconstruction; Jim Crow laws soon disenfranchised many African Americans. In the North, urbanization and an unprecedented influx of immigrants from Southern Southern and Eastern Europe hastened the country's industrialization. The wave of immigration, lasting until 1929, provided labor and transformed American culture. National infrastructure development spurred economic growth. The 1867 Alaska Purchase from Russia completed the country's mainland expansion. The Wounded Knee Massacre in 1890 was the last major armed conflict of the

Indian Wars. In 1893, the indigenous monarchy of the Pacific Kingdom of Hawaii was overthrown in a coup led by American residents; the United States annexed the archipelago in 1898. Victory in the Spanish–American War the same year demonstrated that the United States was a world power and led to the annexation of Puerto Rico, Guam, and the Philippines.[42] The Philippines gained independence a half-century later; Puerto Rico and Guam remain U.S. territories.

World War I, Great Depression, and World War II

At the outbreak of World War I in 1914, the United States remained neutral. Most Americans sympathized with the British and French, although many opposed intervention.[43] In 1917, the United States joined the Allies, and the American Expeditionary Forces helped to turn the tide against the Central Powers. After the war, the Senate did not ratify the Treaty of Versailles, which established the League of Nations. The country pursued a policy of unilateralism, verging on isolationism.[44] In 1920, the women's rights movement won passage of a constitutional amendment granting women's suffrage. The prosperity of the Roaring Twenties ended with the Wall Street Crash of 1929 that triggered the Great Depression. After his election as president in 1932, Franklin D. Roosevelt responded with the New Deal, a range of policies increasing government intervention in the economy, including the establishment of the Social Security system.[45] The Dust Bowl of the mid-1930s impoverished many farming communities and spurred a new wave of western migration.

Soldiers of the U.S. Army 1st Infantry Division landing in Normandy on D-Day, June 6, 1944

The United States, effectively neutral during World War II's early stages after Nazi Germany's invasion of Poland in September 1939, began supplying materiel to the Allies in March 1941 through the Lend-Lease program. On December 7, 1941, the Empire of Japan launched a surprise attack on Pearl Harbor, prompting the United States to join the Allies against the Axis powers as well as the internment of Japanese Americans by the thousands.[46] Participation in the war spurred capital investment and industrial capacity. Among the major combatants, the United States was the only nation to become richer—indeed, far richer—instead of poorer because of the war.[47] Allied conferences at Bretton Woods and Yalta outlined a new system of international organizations that placed the United States and Soviet Union at the center of world affairs. As victory was won in Europe, a 1945 international conference held in San Francisco produced the United Nations Charter, which became active after the war.[48] The United States, having developed the first nuclear weapons, used them on the Japanese cities of Hiroshima and Nagasaki in August. Japan surrendered on September 2, ending the war.[49]

Cold War and protest politics

The United States and Soviet Union jockeyed for power after World War II during the Cold War, dominating the military affairs of Europe through NATO and the Warsaw Pact, respectively. While they engaged in proxy wars and developed powerful nuclear arsenals, the two countries avoided direct military conflict. Resisting leftist land and income redistribution projects around the world, the United States often supported authoritarian governments. American troops fought Communist Chinese forces in the Korean War of 1950–53. The House Un-American Activities Committee pursued a series of investigations into suspected leftist subversion, while Senator Joseph McCarthy became the figurehead of anticommunist sentiment.

The 1961 Soviet launch of the first manned spaceflight prompted President John F. Kennedy's call for the United States to be first to land "a man on the moon", achieved in 1969. Kennedy also faced a tense nuclear showdown with Soviet forces in Cuba. Meanwhile, the United States experienced sustained economic expansion. A growing civil rights movement, symbolized and led by African Americans such as Rosa Parks, Martin Luther King, Jr., and James

Bevel, used nonviolence to confront segregation and discrimination. Following Kennedy's assassination in 1963, the Civil Rights Act of 1964 and Voting Rights Act of 1965 were passed under President Lyndon B. Johnson. He also signed into law the Medicare and Medicaid programs. Johnson and his successor, Richard Nixon, expanded a proxy war in Southeast Asia into the unsuccessful Vietnam War. A widespread countercultural movement grew, fueled by opposition to the war, black nationalism, and the sexual revolution. Betty Friedan, Gloria Steinem, and others led a new wave of feminism that sought political, social, and economic equality for women. As a result of the Watergate scandal, in 1974 Nixon became the first U.S. president to resign, to avoid being impeached on charges including obstruction of justice and abuse of power. The Jimmy Carter administration of the late 1970s was marked by stagflation and the Iran hostage crisis. The election of Ronald Reagan as president in 1980 heralded a rightward shift in American politics, reflected in major changes in taxation and spending priorities. His second term in office brought both the Iran-Contra scandal and significant diplomatic progress with the Soviet Union. The subsequent Soviet collapse ended the Cold War.

Contemporary era

The World Trade Center on the morning of September 11, 2001

Under President George H. W. Bush, the United States took a lead role in the UN–sanctioned Gulf War. The longest economic expansion in modern U.S. history—from March 1991 to March 2001—encompassed the Bill Clinton administration and the dot-com bubble.[50] A civil lawsuit and sex scandal led to Clinton's impeachment in 1998, but he remained in office. The 2000 presidential election, one of the closest in American history, was resolved by a U.S. Supreme Court decision—George W. Bush, son of George H. W. Bush, became president.

On September 11, 2001, al-Qaeda terrorists struck the World Trade Center in New York City and The Pentagon near Washington, D.C., killing nearly three thousand people. In response, the Bush administration launched the global War on Terror, invading Afghanistan and removing the Taliban government and al-Qaeda training camps. Taliban insurgents continue to fight a guerrilla war. In 2002, the Bush administration began to press for regime change in Iraq on controversial grounds.[51] [52] Forces of a so-called Coalition of the Willing invaded Iraq in 2003, ousting Saddam Hussein. In 2005, Hurricane Katrina caused severe destruction along much of the Gulf Coast, devastating New Orleans. In 2008, amid a global economic recession, the first African American president, Barack Obama, was elected. In 2010, major health care and financial system reforms were enacted.

Government, elections, and politics

The west front of the United States Capitol, which houses the United States Congress

The United States is the world's oldest surviving federation. It is a constitutional republic and representative democracy, "in which majority rule is tempered by minority rights protected by law".[53] The government is regulated by a system of checks and balances defined by the U.S. Constitution, which serves as the country's supreme legal document. In the American federalist system, citizens are usually subject to three levels of government, federal, state, and local; the local government's duties are commonly split between county and municipal governments. In almost all cases, executive and legislative officials are elected by a plurality vote of citizens by district. There is no proportional representation at the federal level, and it is very rare at lower levels.

The south façade of the White House, home and workplace of the U.S. president

The federal government is composed of three branches:

- Legislative: The bicameral Congress, made up of the Senate and the House of Representatives, makes federal law, declares war, approves treaties, has the power of the purse, and has the power of impeachment, by which it can remove sitting members of the government.
- Executive: The president is the commander-in-chief of the military, can veto legislative bills before they become law, and appoints the members of the Cabinet (subject to Senate approval) and other officers, who administer and enforce federal laws and policies.
- Judicial: The Supreme Court and lower federal courts, whose judges are appointed by the president with Senate approval, interpret laws and overturn those they find unconstitutional.

The west front of the United States Supreme Court Building

The House of Representatives has 435 voting members, each representing a congressional district for a two-year term. House seats are apportioned among the states by population every tenth year. As of the 2000 census, seven states have the minimum of one representative, while California, the most populous state, has fifty-three. The Senate has 100 members with each state having two senators, elected at-large to six-year terms; one third of Senate seats are up for election every other year. The president serves a four-year term and may be elected to the office no more than twice. The president is not elected by direct vote, but by an indirect electoral college system in which the determining votes are apportioned to the states and the District of Columbia. The Supreme Court, led by the Chief Justice of the United States, has nine members, who serve for life.

The state governments are structured in roughly similar fashion; Nebraska uniquely has a unicameral legislature. The governor (chief executive) of each state is directly elected. Some state judges and cabinet officers are appointed by the governors of the respective states, while others are elected by popular vote.

The original text of the Constitution establishes the structure and responsibilities of the federal government and its relationship with the individual states. Article One protects the right to the "great writ" of habeas corpus, and Article Three guarantees the right to a jury trial in all criminal cases. Amendments to the Constitution require the approval of three-fourths of the states. The Constitution has been amended twenty-seven times; the first ten amendments, which make up the Bill of Rights, and the Fourteenth Amendment form the central basis of Americans' individual rights. All laws and governmental procedures are subject to judicial review and any law ruled in violation of the Constitution is voided. The principle of judicial review, not explicitly mentioned in the Constitution, was declared by the Supreme Court in *Marbury v. Madison* (1803).

Parties and ideology

The United States has operated under a two-party system for most of its history. For elective offices at most levels, state-administered primary elections choose the major party nominees for subsequent general elections. Since the general election of 1856, the major parties have been the Democratic Party, founded in 1824, and the Republican Party, founded in 1854. Since the Civil War, only one third-party presidential candidate—former president Theodore Roosevelt, running as a Progressive in 1912—has won as much as 20% of the popular vote.

Within American political culture, the Republican Party is considered center-right or conservative and the Democratic Party is considered center-left or liberal. The states of the Northeast and West Coast and some of the Great Lakes states, known as "blue states", are relatively liberal. The "red states" of the South and parts of the Great

Plains and Rocky Mountains are relatively conservative.

The winner of the 2008 presidential election, Democrat Barack Obama, is the 44th U.S. president. The 2010 midterm elections saw the Republican Party take control of the House and make gains in the Senate, where the Democrats retain the majority. In the 112th United States Congress, the Senate comprises 51 Democrats, two independents who caucus with the Democrats, and 47 Republicans; the House comprises 240 Republicans and 192 Democrats—three seats are vacant. There are 29 Republican and 20 Democratic state governors, as well as one independent.

Foreign relations and military

The United States exercises global economic, political, and military influence. It is a permanent member of the United Nations Security Council and New York City hosts the United Nations Headquarters. It is a member of the G8, G20, and Organisation for Economic Co-operation and Development. Almost all countries have embassies in Washington, D.C., and many have consulates around the country. Likewise, nearly all nations host American diplomatic missions. However, Cuba, Iran, North Korea, Bhutan, Libya, and the Republic of China (Taiwan) do not have formal diplomatic relations with the United States.

British Foreign Secretary William Hague and U.S. Secretary of State Hillary Clinton, May 2010

The United States has a "special relationship" with the United Kingdom[54] and strong ties with Canada,[55] Australia,[56] New Zealand,[57] the Philippines,[58] Japan,[59] South Korea,[60] Israel,[61] and several European countries. It works closely with fellow NATO members on military and security issues and with its neighbors through the Organization of American States and free trade agreements such as the trilateral North American Free Trade Agreement with Canada and Mexico. In 2008, the United States spent a net $25.4 billion on official development assistance, the most in the world. As a share of America's large gross national income (GNI), however, the U.S. contribution of 0.18% ranked last among twenty-two donor states. By contrast, private overseas giving by Americans is relatively generous.[62]

The USS *Abraham Lincoln* aircraft carrier

The president holds the title of commander-in-chief of the nation's armed forces and appoints its leaders, the secretary of defense and the Joint Chiefs of Staff. The United States Department of Defense administers the armed forces, including the Army, Navy, Marine Corps, and Air Force. The Coast Guard is run by the Department of Homeland Security in peacetime and the Department of the Navy in time of war. In 2008, the armed forces had 1.4 million personnel on active duty. The Reserves and National Guard brought the total number of troops to 2.3 million. The Department of Defense also employed about 700,000 civilians, not including contractors.[63]

Military service is voluntary, though conscription may occur in wartime through the Selective Service System. American forces can be rapidly deployed by the Air Force's large fleet of transport aircraft, the Navy's eleven active aircraft carriers, and Marine Expeditionary Units at sea with the Navy's Atlantic and Pacific fleets. The military operates 865 bases and facilities abroad,[64] and maintains deployments greater than 100 active duty personnel in 25 foreign countries.[65] The extent of this global military presence has prompted some scholars to describe the United States as maintaining an "empire of bases".[66]

Total U.S. military spending in 2008, more than $600 billion, was over 41% of global military spending and greater than the next fourteen largest national military expenditures combined. The per capita spending of $1,967 was about nine times the world average; at 4% of GDP, the rate was the second-highest among the top fifteen military spenders, after Saudi Arabia.[67] The proposed base Department of Defense budget for 2012, $553 billion, is a 4.2% increase over 2011; an additional $118 billion is proposed for the military campaigns in Iraq and Afghanistan.[68] As of September 2010, the United States is scheduled to have 96,000 troops deployed to Afghanistan, and 50,000 to Iraq.[69] As of July 25, 2011, the United States had suffered 4,474 military fatalities during the Iraq War,[70] and 1,680 during the War in Afghanistan.[71]

Economy

Economic indicators		
Unemployment	9.1% (September 2011)	[72]
GDP growth	1.3% (2Q 2011), 2.9% (2010)	[73]
CPI inflation	3.8% (August 2010 – August 2011)	[74]
Poverty	15.1% (2010)	[75]
Public debt	$14.70 trillion (September 15, 2011)	[76]
Household net worth	$58.1 trillion (1Q 2011)	[77]

The United States has a capitalist mixed economy, which is fueled by abundant natural resources, a well-developed infrastructure, and high productivity.[78] According to the International Monetary Fund, the U.S. GDP of $15 trillion constitutes 23% of the gross world product at market exchange rates and over 20% of the gross world product at purchasing power parity (PPP).[4] Though larger than any other nation's, its national GDP is about 5% smaller than the GDP of the European Union at PPP in 2008. The country ranks ninth in the world in nominal GDP per capita and sixth in GDP per capita at PPP.[4] The U.S. dollar is the world's primary reserve currency.

The United States is the largest importer of goods and third largest exporter, though exports per capita are relatively low. In 2010, the total U.S. trade deficit was $634.9 billion.[79] Canada, China, Mexico, Japan, and Germany are its top trading partners.[80] In 2010, oil was the largest import commodity, while transportation equipment was the country's largest export.[79] China is the largest foreign holder of U.S. public debt.[81]

Wall Street and the New York Stock Exchange, the world's largest bourse by dollar volume[82]

In 2009, the private sector was estimated to constitute 55.3% of the economy, with federal government activity accounting for 24.1% and state and local government activity (including federal transfers) the remaining 20.6%.[83] While its economy has reached a postindustrial level of development and its service sector constitutes 67.8% of GDP, the United States remains an industrial power.[84] The leading business field by gross business receipts is wholesale and retail trade; by net income it is manufacturing.[85] Chemical products are the leading manufacturing field.[86] The United States is the third largest producer of oil in the world, as well as its largest importer.[87] It is the world's number one producer of electrical and nuclear energy, as well as liquid natural gas, sulfur, phosphates, and salt. While agriculture accounts for just under 1% of GDP,[84] the United States is the world's top producer of corn[88] and soybeans.[89] Coca-Cola and McDonald's are the two most recognized brands in the world.[90]

In August 2010, the American labor force comprised 154.1 million people. With 21.2 million people, government is the leading field of employment. The largest private employment sector is health care and social assistance, with

16.4 million people. About 12% of workers are unionized, compared to 30% in Western Europe.[91] The World Bank ranks the United States first in the ease of hiring and firing workers.[92] In 2009, the United States had the third highest labor productivity per person in the world, behind Luxembourg and Norway. It was fourth in productivity per hour, behind those two countries and the Netherlands.[93] Compared to Europe, U.S. property and corporate income tax rates are generally higher, while labor and, particularly, consumption tax rates are lower.[94]

Income and human development

According to the United States Census Bureau, the pretax median household income in 2010 was $49,445. The median ranged from $64,308 among Asian American households to $32,068 among African American households.[75] Using purchasing power parity exchange rates, the overall median is similar to the most affluent cluster of developed nations. After declining sharply during the middle of the 20th century, poverty rates have plateaued since the early 1970s, with 11–15% of Americans below the poverty line every year, and 58.5% spending at least one year in poverty between the ages of 25 and 75.[95] [96] In 2010, 46.2 million Americans lived in poverty, a figure that rose for the fourth year in a row.[75]

A middle-class single-family home in suburban Salinas, California

The U.S. welfare state is one of the least extensive in the developed world, reducing both relative poverty and absolute poverty by considerably less than the mean for rich nations,[97] [98] though combined private and public social expenditures per capita are relatively high.[99] While the American welfare state effectively reduces poverty among the elderly,[100] it provides relatively little assistance to the young.[101] A 2007 UNICEF study of children's well-being in twenty-one industrialized nations ranked the United States next to last.[102]

Between 1947 and 1979, real median income rose by over 80% for all classes, with the incomes of poor Americans rising faster than those of the rich.[103] However, income gains since then have been slower, less widely shared, and accompanied by increased economic insecurity.[103] [104] Median household income has increased for all classes since 1980,[105] largely owing to more dual-earner households, the closing of the gender gap, and longer work hours, but the growth has been strongly tilted toward the very top.[97] [103] [106] Consequently, the share of income of the top 1%—21.8% of total reported income in 2005—has more than doubled since 1980,[107] leaving the United States with the greatest income inequality among developed nations.[97] [108] The top 1% pays 27.6% of all federal taxes, while the top 10% pays 54.7%.[109] Wealth, like income, is highly concentrated: The richest 10% of the adult population possesses 69.8% of the country's household wealth, the second-highest share among developed nations.[110] The top 1% possesses 33.4% of net wealth.[111] In 2010 the United Nations Development Programme ranked the United States 12th among 139 countries on its inequality-adjusted human development index (IHDI), eight places lower than in the standard HDI.[112]

Infrastructure

Science and technology

A photograph from Apollo 11 of Buzz Aldrin on the surface of the Moon

The United States has been a leader in scientific research and technological innovation since the late 19th century. In 1876, Alexander Graham Bell was awarded the first U.S. patent for the telephone. Thomas Edison's laboratory developed the phonograph, the first long-lasting light bulb, and the first viable movie camera. Nikola Tesla pioneered alternating current, the AC motor, and radio. In the early 20th century, the automobile companies of Ransom E. Olds and Henry Ford popularized the assembly line. The Wright brothers, in 1903, made the first sustained and controlled heavier-than-air powered flight.[113]

The rise of Nazism in the 1930s led many European scientists, including Albert Einstein and Enrico Fermi, to immigrate to the United States. During World War II, the Manhattan Project developed nuclear weapons, ushering in the Atomic Age. The Space Race produced rapid advances in rocketry, materials science, and computers. IBM, Apple Computer, and Microsoft refined and popularized the personal computer. The United States largely developed the ARPANET and its successor, the Internet. Today, 64% of research and development funding comes from the private sector.[114] The United States leads the world in scientific research papers and impact factor.[115] Americans possess high levels of technological consumer goods,[116] and almost half of U.S. households have broadband Internet access.[117] The country is the primary developer and grower of genetically modified food, representing half of the world's biotech crops.[118]

Transportation

The Interstate Highway System, which extends 46876 miles (75440 km)[119]

Personal transportation is dominated by automobiles, which operate on a network of 13 million roads,[120] including the world's longest highway system.[121] The world's second largest automobile market,[122] the United States has the highest rate of per-capita vehicle ownership in the world, with 765 vehicles per 1,000 Americans.[123] About 40% of personal vehicles are vans, SUVs, or light trucks.[124] The average American adult (accounting for all drivers and nondrivers) spends 55 minutes driving every day, traveling 29 miles (47 km).[125]

Mass transit accounts for 9% of total U.S. work trips,[126] ranking last in a survey of 17 countries.[127] While transport of goods by rail is extensive, relatively few people use rail to travel.[128] Light rail development has increased in recent years but, like high speed rail, is below European levels.[129] Bicycle usage for work commutes is minimal.[130]

The civil airline industry is entirely privately owned and has been largely deregulated since 1978, while most major airports are publicly owned. The four largest airlines in the world by passengers carried are American; Southwest Airlines is number one.[131] Of the world's thirty busiest passenger airports, sixteen are in the United States, including the busiest, Hartsfield-Jackson Atlanta International Airport.[132]

Energy

The United States energy market is 29,000 terawatt hours per year. Energy consumption per capita is 7.8 tons of oil equivalent per year, the 10th highest rate in the world. In 2005, 40% of this energy came from petroleum, 23% from coal, and 22% from natural gas. The remainder was supplied by nuclear power and renewable energy sources.[133] The United States is the world's largest consumer of petroleum.[134] For decades, nuclear power has played a limited role relative to many other developed countries, in part due to public perception in the wake of a 1979 accident. In 2007, several applications for new nuclear plants were filed.[135] The United States has 27% of global coal reserves.[136]

Education

Harvard University, the country's oldest institution of higher learning and its first corporation, is one of the most prestigious universities in the world.[137]

American public education is operated by state and local governments, regulated by the United States Department of Education through restrictions on federal grants. Children are required in most states to attend school from the age of six or seven (generally, kindergarten or first grade) until they turn eighteen (generally bringing them through twelfth grade, the end of high school); some states allow students to leave school at sixteen or seventeen.[138] About 12% of children are enrolled in parochial or nonsectarian private schools. Just over 2% of children are homeschooled.[139]

The United States has many competitive private and public institutions of higher education. According to prominent international rankings, 13 or 15 American colleges and universities are ranked among the top 20 in the world.[140][141] There are also local community colleges with generally more open admission policies, shorter academic programs, and lower tuition. Of Americans twenty-five and older, 84.6% graduated from high school, 52.6% attended some college, 27.2% earned a bachelor's degree, and 9.6% earned graduate degrees.[142] The basic literacy rate is approximately 99%.[1] [143] The United Nations assigns the United States an Education Index of 0.97, tying it for 12th in the world.[144]

Health

The United States life expectancy of 78.3 years at birth is ranked 36th among 194 United Nations member states; while above the world average, it falls short of the overall figure in Western Europe.[145] Increasing obesity in the United States and health improvements elsewhere have contributed to lowering the country's rank in life expectancy from 1987 to 2007, from 11th to 42nd in the world.[146] The infant mortality rate of 6.37 per thousand places the United States 42nd out of 221 countries, above average but behind all of Western Europe.[147] Approximately one-third of the adult population is obese and an additional third is overweight;[148] the obesity rate, the highest in the industrialized world, has more than doubled in the last quarter-century.[149] Obesity-related type 2 diabetes is considered epidemic by health care professionals.[150]

The U.S. health care system far outspends any other nation's, measured in both per capita spending and percentage of GDP.[152] The World Health Organization ranked the U.S. health care system in 2000 as first in responsiveness, but 37th in overall performance.

The Texas Medical Center in Houston, the world's largest medical center[151]

Health care coverage in the United States is a combination of public and private efforts, and is not universal as in all other developed countries. In 2004, private insurance paid for 36% of personal health expenditures, private out-of-pocket payments covered 15%, and federal, state, and local governments paid for 44%.[153] In 2005, 46.6 million Americans, 15.9% of the population, were uninsured, 5.4 million more than in 2001. The main cause of this rise is the drop in the number of Americans with employer-sponsored health insurance.[154] The subject of uninsured and underinsured Americans is a major political issue.[155] A 2009 study estimated that lack of insurance is associated with nearly 45,000 deaths a year.[156] In 2006, Massachusetts became the first state to mandate universal health insurance.[157] Federal legislation passed in early 2010 will create a near-universal health insurance system around the country by 2014.

Crime and law enforcement

Law enforcement in the United States is primarily the responsibility of local police and sheriff's departments, with state police providing broader services. Federal agencies such as the Federal Bureau of Investigation (FBI) and the U.S. Marshals Service have specialized duties. At the federal level and in almost every state, jurisprudence operates on a common law system. State courts conduct most criminal trials; federal courts handle certain designated crimes as well as certain appeals from the state systems. Federal law prohibits a variety of drugs, although states sometimes pass laws in conflict with federal regulations. The smoking age is generally 18 and the drinking age is generally 21. Among developed nations, the United States has above-average levels of violent crime and particularly high levels of gun violence and homicide.[158] There were 5.0 murders per 100,000 persons in 2009, 10.4% fewer than in 2000.[159] Gun ownership rights are the subject of contentious political debate.

The United States has the highest documented incarceration rate[160] and total prison population[161] in the world. At the start of 2008, more than 2.3 million people were incarcerated, more than one in every 100 adults.[162] The current rate is about seven times the 1980 figure,[163] and over three times the figure in Poland, the Organisation for Economic Co-operation and Development (OECD) country with the next highest rate.[164] African American males are jailed at about six times the rate of white males and three times the rate of Hispanic males.[160] The country's high rate of incarceration is largely due to sentencing and drug policies.[160] [165]

Though it has been abolished in most Western nations, capital punishment is sanctioned in the United States for certain federal and military crimes, and in thirty-four states. Since 1976, when the U.S. Supreme Court reinstated the death penalty after a four-year moratorium, there have been more than 1,000 executions.[166] In 2010, the country had the fifth highest number of executions in the world, following China, Iran, North Korea, and Yemen.[167] In 2007, New Jersey became the first state to legislatively abolish the death penalty since the 1976 Supreme Court decision, followed by New Mexico in 2009 and Illinois in 2011.[168]

Demographics

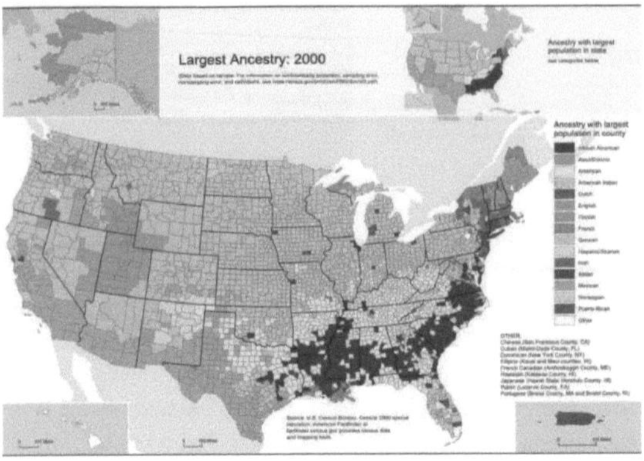

Largest ancestry groups by county, 2000

Race/Ethnicity (2010)[169]	
White	72.4%
Black/African American	12.6%
Asian	4.8%
American Indian and Alaska Native	0.9%
Native Hawaiian and Pacific Islander	0.2%
Other	6.2%
Two or more races	2.9%
Hispanic/Latino (*of any race*)	16.3%

The U.S. Census Bureau estimates the country's population now to be 312421000,[2] including an estimated 11.2 million illegal immigrants.[170] The U.S. population almost quadrupled during the 20th century, from about 76 million in 1900.[171] The third most populous nation in the world, after China and India, the United States is the only industrialized nation in which large population increases are projected.[172] Even with a birth rate of 13.82 per 1,000, 30% below the world average, its population growth rate is positive at 1%, significantly higher than those of many developed nations.[173] In fiscal year 2010, over 1 million immigrants (most of whom entered through family reunification) were granted legal residence.[174] Mexico has been the leading source of new residents for over two decades; since 1998, China, India, and the Philippines have been in the top four sending countries every year.[175]

The United States has a very diverse population—thirty-one ancestry groups have more than one million members.[176] White Americans are the largest racial group; German Americans, Irish Americans, and English Americans constitute three of the country's four largest ancestry groups.[176] African Americans are the nation's largest racial minority and third largest ancestry group.[176] Asian Americans are the country's second largest racial minority; the two largest Asian American ethnic groups are Chinese Americans and Filipino Americans.[176] In 2010, the U.S. population included an estimated 5.2 million people with some American Indian or Alaska Native ancestry (2.9 million exclusively of such ancestry) and 1.2 million with some native Hawaiian or Pacific island ancestry (0.5 million exclusively).[177] The census counted more than 19 million people of "Some Other Race" who were "unable to identify with any" of its five official race categories in 2010.[177]

The population growth of Hispanic and Latino Americans (the terms are officially interchangeable) is a major demographic trend. The 50.5 million Americans of Hispanic descent[177] are identified as sharing a distinct "ethnicity" by the Census Bureau; 64% of Hispanic Americans are of Mexican descent.[178] Between 2000 and 2010, the country's Hispanic population increased 43% while the non-Hispanic population rose just 4.9%.[169] Much of this growth is from immigration; as of 2007, 12.6% of the U.S. population was foreign-born, with 54% of that figure born in Latin America.[179] Fertility is also a factor; the average Hispanic woman gives birth to 3.0 children in her lifetime, compared to 2.2 for non-Hispanic black women and 1.8 for non-Hispanic white women (below the replacement rate of 2.1).[172] Minorities (as defined by the Census Bureau as all those beside non-Hispanic, non-multiracial whites) constitute 36.3% of the population in 2010,[180] and are projected to constitute the majority by 2042.[181]

About 82% of Americans live in urban areas (including suburbs);[1] about half of those reside in cities with populations over 50,000.[182] In 2008, 273 incorporated places had populations over 100,000, nine cities had more than 1 million residents, and four global cities had over 2 million (New York City, Los Angeles, Chicago, and Houston).[183] There are fifty-two metropolitan areas with populations greater than 1 million.[184] Of the fifty fastest-growing metro areas, forty-seven are in the West or South.[185] The metro areas of Dallas, Houston, Atlanta, and Phoenix all grew by more than a million people between 2000 and 2008.[184]

Leading population centers

Rank	Core City	Metro area pop.[186]	Metropolitan Statistical Area	Region[187]
1	**New York**	18,897,109	New York-Northern New Jersey-Long Island, NY-NJ-PA MSA	Northeast
2	**Los Angeles**	12,828,837	Los Angeles-Long Beach-Santa Ana, CA MSA	West
3	**Chicago**	9,461,105	Chicago-Naperville-Joliet, IL-IN-WI MSA	Midwest
4	**Dallas**	6,371,773	Dallas-Fort Worth-Arlington, TX MSA	South
5	**Philadelphia**	5,965,343	Philadelphia-Camden-Wilmington, PA-NJ-DE-MD MSA	Northeast
6	**Houston**	5,946,800	Houston-Sugar Land-Baytown, TX MSA	South
7	**Washington, D.C.**	5,582,170	Washington-Arlington-Alexandria, DC-VA-MD-WV MSA	South
8	**Miami**	5,564,635	Miami-Fort Lauderdale-Pompano Beach, FL MSA	South
9	**Atlanta**	5,268,860	Atlanta-Sandy Springs-Marietta, GA MSA	South
10	**Boston**	4,552,402	Boston-Cambridge-Quincy, MA-NH MSA	Northeast

based on the 2010 U.S. Census

New York

Los Angeles

Language

Languages (2007)[188]	
English (*only*)	225.5 million
Spanish, incl. Creole	34.5 million
Chinese	2.5 million
French, incl. Creole	2.0 million
Tagalog	1.5 million
Vietnamese	1.2 million
German	1.1 million
Korean	1.1 million

English is the de facto national language. Although there is no official language at the federal level, some laws—such as U.S. naturalization requirements—standardize English. In 2007, about 226 million, or 80% of the population aged five years and older, spoke only English at home. Spanish, spoken by 12% of the population at home, is the second most common language and the most widely taught second language.[188] [189] Some Americans advocate making English the country's official language, as it is in at least twenty-eight states.[6] Both Hawaiian and English are official languages in Hawaii by state law.[190]

While neither has an official language, New Mexico has laws providing for the use of both English and Spanish, as Louisiana does for English and French.[191] Other states, such as California, mandate the publication of Spanish versions of certain government documents including court forms.[192] Many jurisdictions with large numbers of non-English speakers produce government materials, especially voting information, in the most commonly spoken languages in those jurisdictions. Several insular territories grant official recognition to their native languages, along with English: Samoan and Chamorro are recognized by American Samoa and Guam, respectively; Carolinian and Chamorro are recognized by the Northern Mariana Islands; Spanish is an official language of Puerto Rico.

Religion

A Presbyterian church; most Americans identify as Christian.

The United States is officially a secular nation; the First Amendment of the U.S. Constitution guarantees the free exercise of religion and forbids the establishment of any religious governance. In a 2002 study, 59% of Americans said that religion played a "very important role in their lives", a far higher figure than that of any other wealthy nation.[193] According to a 2007 survey, 78.4% of adults identified themselves as Christian,[194] down from 86.4% in 1990.[195] Protestant denominations accounted for 51.3%, while Roman Catholicism, at 23.9%, was the largest individual denomination. The study categorizes white evangelicals, 26.3% of the population, as the country's largest religious cohort;[194] another study estimates evangelicals of all races at 30–35%.[196] The total reporting non-Christian religions in 2007 was 4.7%, up from 3.3% in 1990.[195] The leading non-Christian faiths were Judaism (1.7%), Buddhism (0.7%), Islam (0.6%), Hinduism (0.4%), and Unitarian Universalism (0.3%).[194] The survey also reported that 16.1% of Americans described themselves as agnostic, atheist, or simply having no religion, up from 8.2% in 1990.[194] [195]

Family structure

In 2007, 58% of Americans age 18 and over were married, 6% were widowed, 10% were divorced, and 25% had never been married.[197] Women now mostly work outside the home and receive a majority of bachelor's degrees.[198]

Same-sex marriage is a contentious issue. Some states permit civil unions or domestic partnerships in lieu of marriage. Since 2003, several states have legalized gay marriage as the result of judicial or legislative action. Meanwhile, the federal government and a majority of states define marriage as between a man and a woman and/or explicitly prohibit same-sex marriage. Public opinion on the issue has shifted from general opposition in the 1990s to a statistical deadlock as of 2011.[199]

The U.S. teenage pregnancy rate, 79.8 per 1,000 women, is the highest among OECD nations.[200] Abortion policy was left to the states until the Supreme Court legalized the practice in 1972. The issue remains highly controversial, with public opinion closely divided for many years. Many states ban public funding of the procedure and restrict late-term abortions, require parental notification for minors, and mandate a waiting period. While the abortion rate is falling, the abortion ratio of 241 per 1,000 live births and abortion rate of 15 per 1,000 women aged 15–44 remain higher than those of most Western nations.[201]

Culture

The United States is a multicultural nation, home to a wide variety of ethnic groups, traditions, and values.[7] [202] Aside from the now small Native American and Native Hawaiian populations, nearly all Americans or their ancestors immigrated within the past five centuries.[203] The culture held in common by most Americans—mainstream American culture—is a Western culture largely derived from the traditions of European immigrants with influences from many other sources, such as traditions brought by slaves from Africa.[7] [204] More recent immigration from Asia and especially Latin America has added to a cultural mix that has been described as both a homogenizing melting pot, and a heterogeneous salad bowl in which immigrants and their descendants retain distinctive cultural characteristics.[7]

American cultural icons: apple pie, baseball, and the American flag

American culture is considered the most individualistic in the world.[205] Though the American Dream, or the perception that Americans enjoy high social mobility, plays a key role in attracting immigrants, other developed nations offer greater social mobility.[206] While the mainstream culture holds that the United States is a classless society,[207] scholars identify significant differences between the country's social classes, affecting socialization, language, and values.[208] The American middle and professional class has initiated many contemporary social trends such as modern feminism, environmentalism, and multiculturalism.[209] Americans' self-images, social viewpoints, and cultural expectations are associated with their occupations to an unusually close degree.[210] While Americans tend greatly to value socioeconomic achievement, being ordinary or average is generally seen as a positive attribute.[211]

Popular media

The Hollywood Sign

The world's first commercial motion picture exhibition was given in New York City in 1894, using Thomas Edison's Kinetoscope. The next year saw the first commercial screening of a projected film, also in New York, and the United States was in the forefront of sound film's development in the following decades. Since the early 20th century, the U.S. film industry has largely been based in and around Hollywood, California. Director D. W. Griffith was central to the development of film grammar and Orson Welles's *Citizen Kane* (1941) is frequently cited as the greatest film of all time.[212] American screen actors like John Wayne and Marilyn Monroe have become iconic figures, while producer/entrepreneur Walt Disney was a leader in both animated film and movie merchandising. The major film studios of Hollywood have produced the most commercially successful movies in history, such as *Star Wars* (1977) and *Titanic* (1997), and the products of Hollywood today dominate the global film industry.[213]

Americans are the heaviest television viewers in the world,[214] and the average viewing time continues to rise, reaching five hours a day in 2006.[215] The four major broadcast networks are all commercial entities. Americans listen to radio programming, also largely commercialized, on average just over two-and-a-half hours a day.[216] Aside from web portals and search engines, the most popular websites are Facebook, YouTube, Wikipedia, Blogger, eBay, and Craigslist.[217]

The rhythmic and lyrical styles of African-American music have deeply influenced American music at large, distinguishing it from European traditions. Elements from folk idioms such as the blues and what is now known as old-time music were adopted and transformed into popular genres with global audiences. Jazz was developed by innovators such as Louis Armstrong and Duke Ellington early in the 20th century. Country music developed in the 1920s, and rhythm and blues in the 1940s. Elvis Presley and Chuck Berry were among the mid-1950s pioneers of rock and roll. In the 1960s, Bob Dylan emerged from the folk revival to become one of America's most celebrated songwriters and James Brown led the development of funk. More recent American creations include hip hop and house music. American pop stars such as Presley, Michael Jackson, and Madonna have become global celebrities.[218]

Literature, philosophy, and the arts

Jack Kerouac, one of the best-known figures of the Beat Generation, a group of writers that came to prominence in the 1950s

In the 18th and early 19th centuries, American art and literature took most of its cues from Europe. Writers such as Nathaniel Hawthorne, Edgar Allan Poe, and Henry David Thoreau established a distinctive American literary voice by the middle of the 19th century. Mark Twain and poet Walt Whitman were major figures in the century's second half; Emily Dickinson, virtually unknown during her lifetime, is now recognized as an essential American poet.[219] A work seen as capturing fundamental aspects of the national experience and character—such as Herman Melville's *Moby-Dick* (1851), Twain's *The Adventures of Huckleberry Finn* (1885), and F. Scott Fitzgerald's *The Great Gatsby* (1925)—may be dubbed the "Great American Novel".[220]

Eleven U.S. citizens have won the Nobel Prize in Literature, most recently Toni Morrison in 1993. William Faulkner and Ernest Hemingway are often named among the most influential writers of the 20th century.[221] Popular literary genres such as the Western and hardboiled crime fiction developed in the United States. The Beat Generation writers opened up new literary approaches, as have postmodernist authors such as John Barth, Thomas Pynchon, and Don DeLillo.

The transcendentalists, led by Thoreau and Ralph Waldo Emerson, established the first major American philosophical movement. After the Civil War, Charles Sanders Peirce and then William James and John Dewey were leaders in the development of pragmatism. In the 20th century, the work of W. V. Quine and Richard Rorty, built upon by Noam Chomsky, brought analytic philosophy to the fore of U.S. academics. John Rawls and Robert Nozick led a revival of political philosophy.

In the visual arts, the Hudson River School was a mid-19th-century movement in the tradition of European naturalism. The realist paintings of Thomas Eakins are now widely celebrated. The 1913 Armory Show in New York City, an exhibition of European modernist art, shocked the public and transformed the U.S. art scene.[222] Georgia O'Keeffe, Marsden Hartley, and others experimented with new styles, displaying a highly individualistic sensibility. Major artistic movements such as the abstract expressionism of Jackson Pollock and Willem de Kooning and the pop art of Andy Warhol and Roy Lichtenstein developed largely in the United States. The tide of modernism and then postmodernism has brought fame to American architects such as Frank Lloyd Wright, Philip Johnson, and Frank Gehry.

Times Square in New York City, part of the Broadway theater district

One of the first major promoters of American theater was impresario P. T. Barnum, who began operating a lower Manhattan entertainment complex in 1841. The team of Harrigan and Hart produced a series of popular musical comedies in New York starting in the late 1870s. In the 20th century, the modern musical form emerged on Broadway; the songs of musical theater composers such as Irving Berlin, Cole Porter, and Stephen Sondheim have become pop standards. Playwright Eugene O'Neill won the Nobel literature prize in 1936; other acclaimed U.S. dramatists include multiple Pulitzer Prize winners Tennessee Williams, Edward Albee, and August Wilson.

Though little known at the time, Charles Ives's work of the 1910s established him as the first major U.S. composer in the classical tradition, while experimentalists such as Henry Cowell and John Cage created a distinctive American approach to classical composition. Aaron Copland and George Gershwin developed a new synthesis of popular and classical music. Choreographers Isadora Duncan and Martha Graham helped create modern dance, while George Balanchine and Jerome Robbins were leaders in 20th century ballet. Americans have long been important in the modern artistic medium of photography, with major photographers including Alfred Stieglitz, Edward Steichen, and Ansel Adams. The newspaper comic strip and the comic book are both U.S. innovations. Superman, the quintessential comic book superhero, has become an American icon.[223]

Food

Mainstream American cuisine is similar to that in other Western countries. Wheat is the primary cereal grain. Traditional American cuisine uses indigenous ingredients, such as turkey, venison, potatoes, sweet potatoes, corn, squash, and maple syrup, which were consumed by Native Americans and early European settlers. Slow-cooked pork and beef barbecue, crab cakes, potato chips, and chocolate chip cookies are distinctively American foods. Soul food, developed by African slaves, is popular around the South and among many African Americans elsewhere. Syncretic cuisines such as Louisiana creole, Cajun, and Tex-Mex are regionally important.

Characteristic dishes such as apple pie, fried chicken, pizza, hamburgers, and hot dogs derive from the recipes of various immigrants. French fries, Mexican dishes such as burritos and tacos, and pasta dishes freely adapted from Italian sources are widely consumed.[224] Americans generally prefer coffee to tea. Marketing by U.S. industries is largely responsible for making orange juice and milk ubiquitous breakfast beverages.[225]

The American fast food industry, the world's largest, pioneered the drive-through format in the 1930s. Fast food consumption has sparked health concerns. During the 1980s and 1990s, Americans' caloric intake rose 24%;[224] frequent dining at fast food outlets is associated with what public health officials call the American "obesity epidemic".[226] Highly sweetened soft drinks are widely popular, and sugared beverages account for 9% of American caloric intake.[227]

Sports

A college football quarterback looking to pass the ball

Baseball has been regarded as the national sport since the late 19th century, while American football is now by several measures the most popular spectator sport.[228] Basketball and ice hockey are the country's next two leading professional team sports. College football and basketball attract large audiences. Boxing and horse racing were once the most watched individual sports, but they have been eclipsed by golf and auto racing, particularly NASCAR. Soccer is played widely at the youth and amateur levels. Tennis and many outdoor sports are popular as well.

While most major U.S. sports have evolved out of European practices, basketball, volleyball, skateboarding, snowboarding, and cheerleading are American inventions. Lacrosse and surfing arose from Native American and Native Hawaiian activities that predate Western contact. Eight Olympic Games have taken place in the United States. The United States has won 2,301 medals at the Summer Olympic Games, more than any other country,[229] and 253 in the Winter Olympic Games, the second most.[230]

Measurement systems

The nation retains United States customary units, comprising mainly former British imperial units such as miles, yards, and degrees Fahrenheit. Distinct units include the U.S. gallon and U.S. pint volume measurements. The United States is one of only three countries that do not rely primarily on the International System of Units. However, metric units are increasingly used in science, medicine, and many industrial fields.[231]

See also

- International rankings of the United States
- Lists of Americans
- List of places in the United States
- Wikipedia:List of U.S. state portals

References

[1] "United States" (https://www.cia.gov/library/publications/the-world-factbook/geos/us.html). *The World Factbook*. CIA. 2009-09-30. . Retrieved 2010-01-05 (area given in square kilometers).

[2] "U.S. POPClock Projection" (http://www.census.gov/population/www/popclockus.html). U.S. Census Bureau. . Figure updated automatically.

[3] "United States" (http://www.imf.org/external/pubs/ft/weo/2011/02/weodata/weorept.aspx?sy=2009&ey=2016&scsm=1&ssd=1&sort=country&ds=.&br=1&c=111&s=NGDPD,NGDPDPC,PPPGDP,PPPPC,LP&grp=0&a=&pr1.x=49&pr1.y=17). International Monetary Fund. . Retrieved 2011-10-09.

[4] "World Economic Outlook Database" (http://www.imf.org/external/pubs/ft/weo/2011/02/weodata/weorept.aspx?sy=2009&ey=2011&scsm=1&ssd=1&sort=country&ds=.&br=1&c=111&s=NGDPD,NGDPDPC,PPPGDP,PPPPC,PPPSH,LP&grp=0&a=&pr.x=35&pr.y=14). International Monetary Fund. September 2011. . Retrieved 2011-09-11.

[5] "Human Development Report 2010" (http://hdr.undp.org/en/media/HDR_2010_EN_Tables_reprint.pdf). United Nations. 2010. . Retrieved 4 November 2010.

[6] Feder, Jody (2007-01-25). "English as the Official Language of the United States—Legal Background and Analysis of Legislation in the 110th Congress" (http://www.ilw.com/immigrationdaily/news/2007,0515-crs.pdf). Ilw.com (Congressional Research Service). . Retrieved 2007-06-19.

[7] Adams, J. Q., and Pearlie Strother-Adams (2001). *Dealing with Diversity*. Chicago: Kendall/Hunt. ISBN 0-7872-8145-X.

[8] The European Union has a larger collective economy, but is not a single nation.

[9] Dull, Jonathan R. (2003). "Diplomacy of the Revolution, to 1783", p. 352, chap. in *A Companion to the American Revolution*, ed. Jack P. Greene and J. R. Pole. Maiden, Mass.: Blackwell, pp. 352–361. ISBN 1-4051-1674-9.

[10] Maddison, Angus (2006). "Historical Statistics for the World Economy" (http://www.ggdc.net/maddison/Historical_Statistics/horizontal-file_09-2008.xls). The Groningen Growth and Development Centre, Economics Department of the University of Groningen. . Retrieved 2008-11-06.

[11] "US, Allies' Share of World Military Spending Shrinking—Study" (http://www.reuters.com/article/2011/07/07/military-spending-usa-idUSN1E7661J620110707). Reuters. 2011-07-07. . Retrieved 2011-08-08.

[12] Cohen, Eliot A. (July/August 2004). "History and the Hyperpower" (http://www.foreignaffairs.com/articles/59919/eliot-a-cohen/history-and-the-hyperpower). *Foreign Affairs*. . Retrieved 2006-07-14. "Country Profile: United States of America" (http://news.bbc.co.uk/2/hi/americas/country_profiles/1217752.stm). BBC News. 2008-04-22. . Retrieved 2008-05-18.

[13] "Cartographer Put 'America' on the Map 500 years Ago" (http://www.usatoday.com/news/nation/2007-04-24-america-turns-500_N.htm?csp=34). *USA Today*. 2007-04-24. . Retrieved 2008-11-30.

[14] "The Charters of Freedom" (http://www.archives.gov/exhibits/charters/charters.html). National Archives. . Retrieved 2007-06-20.

[15] McClure, James (2008-06-12). "A Primer: The 'First Capital' Debate" (http://www.ydr.com/ci_9569289). YDR.com. . Retrieved 2010-07-26.

[16] Wilson, Kenneth G. (1993). *The Columbia Guide to Standard American English*. New York: Columbia University Press, pp. 27–28. ISBN 0-231-06989-8.

[17] Zimmer, Benjamin (2005-11-24). "Life in These, Uh, This United States" (http://itre.cis.upenn.edu/~myl/languagelog/archives/002663.html). University of Pennsylvania—Language Log. . Retrieved 2008-02-22.

[18] Lubowski, Ruben, Marlow Vesterby, and Shawn Bucholtz (2006-07-21). "AREI Chapter 1.1: Land Use" (http://www.ers.usda.gov/publications/arei/eib16/chapter1/1.1/). Economic Research Service. . Retrieved 2009-03-09.

[19] "United States" (http://www.britannica.com/EBchecked/topic/616563/United-States). Encyclopedia Britannica. . Retrieved 2008-03-25 (area given in square miles).

[20] "Population by Sex, Rate of Population Increase, Surface Area and Density" (http://unstats.un.org/unsd/demographic/products/dyb/DYB2005/Table03.pdf). *Demographic Yearbook 2005*. UN Statistics Division. . Retrieved 2008-03-25 (area given in square kilometers).

[21] "World Factbook: Area Country Comparison Table" (http://web.archive.org/web/20080208233209rn_1/education.yahoo.com/reference/factbook/countrycompare/area/3d.html). Yahoo Education. . Retrieved 2007-02-28.

[22] O'Hanlon, Larry. "Supervolcano: What's Under Yellowstone?" (http://dsc.discovery.com/convergence/supervolcano/under/under.html). Discovery Channel. . Retrieved 2007-06-13.

[23] Perkins, Sid (2002-05-11). "Tornado Alley, USA" (http://web.archive.org/web/20070701131631/http://www.sciencenews.org/articles/20020511/bob9.asp). *Science News*. Archived from the original (http://www.sciencenews.org/articles/20020511/bob9.asp) on 2007-07-01. . Retrieved 2006-09-20.

[24] Morin, Nancy. "Vascular Plants of the United States" (http://www.fungaljungal.org/papers/National_Biological_Service.pdf). *Plants*. National Biological Service. . Retrieved 2008-10-27.

[25] "Global Significance of Selected U.S. Native Plant and Animal Species" (http://www.sdi.gov/curtis/TxTab4x1.html). SDI Group. 2001-02-09. . Retrieved 2009-01-20.

[26] "Numbers of Insects (Species and Individuals)" (http://www.si.edu/Encyclopedia_SI/nmnh/buginfo/bugnos.htm). Smithsonian Institution. . Retrieved 2009-01-20.

[27] "National Park Service Announces Addition of Two New Units" (http://home.nps.gov/applications/release/Detail.cfm?ID=639). National Park Service. 2006-02-28. . Retrieved 2006-06-13.

[28] "Federal Land and Buildings Ownership" (http://johnshadegg.house.gov/rsc/Federal Land Ownership--May 2005.pdf). Republican Study Committee. 2005-05-19. . Retrieved 2009-03-09.

[29] See 8 U.S.C. § 1101(a)(36) and 8 U.S.C. § 1101(a)(38) U.S. Federal Code, Inmigration and Nationality Act. 8 U.S.C. § 1101a (http://www.law.cornell.edu/uscode/8/1101a.html)

[30] "Presidential Elections in the United States: A Primer (Page 7)" (http://www.senate.gov/reference/resources/pdf/RL30527.pdf) (PDF). United States Congressional Research Service. 2000-04-17. . Retrieved 2010-01-18.

[31] Raskin, James B. (2003). *Overruling Democracy: The Supreme Court Vs. the American People*. London and New York: Routledge, pp. 36–38. ISBN 0-415-93439-7.

[32] "Peopling of Americas" (http://web.archive.org/web/20071128083459/http://anthropology.si.edu/HumanOrigins/faq/americas.htm). Smithsonian Institution, National Museum of Natural History. June 2004. Archived from the original (http://anthropology.si.edu/HumanOrigins/faq/americas.htm) on 2007-11-28. . Retrieved 2007-06-19.

[33] Meltzer, D.J. (1992). "How Columbus Sickened the New World: Why Were Native Americans So Vulnerable to the Diseases European Settlers Brought With Them?" (http://www.newscientist.com/article/mg13618424.700-how-columbus-sickened-the-new-world-why-were-nativeamericans-so-vulnerable-to-the-diseases-european-settlers-brought-with-them.

html). *New Scientist*: 38. .

[34] Butler, James Davie (October 1896). "British Convicts Shipped to American Colonies" (http://www.dinsdoc.com/butler-1.htm). *American Historical Review 2*. Smithsonian Institution, National Museum of Natural History. . Retrieved 2007-06-21.

[35] Russell, David Lee (2005). *The American Revolution in the Southern Colonies*. Jefferson, N.C., and London: McFarland, p. 12. ISBN 0-7864-0783-2.

[36] Blackburn, Robin (1998). *The Making of New World Slavery: From the Baroque to the Modern, 1492–1800*. London and New York: Verso, p. 460. ISBN 1-85984-195-3.

[37] "Louisiana Purchase" (http://www.nps.gov/archive/jeff/lewisclark2/circa1804/heritage/louisianapurchase/louisianapurchase.htm). *National Parks Services*. . Retrieved 1 March 2011.

[38] Morrison, Michael A. (1999). *Slavery and the American West: The Eclipse of Manifest Destiny and the Coming of the Civil War*. Chapel Hill: University of North Carolina Press, pp. 13–21. ISBN 0-8078-4796-8.

[39] "1860 Census" (http://www2.census.gov/prod2/decennial/documents/1860a-02.pdf). U.S. Census Bureau. . Retrieved 2007-06-10. Page 7 lists a total slave population of 3,953,760.

[40] De Rosa, Marshall L. (1997). *The Politics of Dissolution: The Quest for a National Identity and the American Civil War*. Edison, NJ: Transaction, p. 266. ISBN 1-56000-349-9.

[41] Maris Vinovskis (1990). " *Toward a social history of the American Civil War: exploratory essays* (http://books.google.com/books?id=gySktxKYPGoC&pg=PA6&dq&hl=en#v=onepage&q=&f=false)". Cambridge University Press. p.6. ISBN 0521395593

[42] Gates, John M. (August 1984). "War-Related Deaths in the Philippines" (http://www3.wooster.edu/History/jgates/book-ch3.html). *Pacific Historical Review*. College of Wooster. . Retrieved 2007-09-27.

[43] Foner, Eric, and John A. Garraty (1991). *The Reader's Companion to American History*. New York: Houghton Mifflin, p. 576. ISBN 0-395-51372-3.

[44] McDuffie, Jerome, Gary Wayne Piggrem, and Steven E. Woodworth (2005). *U.S. History Super Review*. Piscataway, NJ: Research & Education Association, p. 418. ISBN 0-7386-0070-9.

[45] Axinn, June; Stern, Mark J. (2007). *Social Welfare: A History of the American Response to Need* (7 ed.). Allyn & Bacon. ISBN 978-0205522156.

[46] Burton, Jeffrey F., et al. (July 2000). "A Brief History of Japanese American Relocation During World War II" (http://www.nps.gov/history/history/online_books/anthropology74/ce3.htm). *Confinement and Ethnicity: An Overview of World War II Japanese American Relocation Sites*. National Park Service. . Retrieved 2010-04-02.

[47] Kennedy, Paul (1989). *The Rise and Fall of the Great Powers*. New York: Vintage, p. 358. ISBN 0-679-72019-7.

[48] "The United States and the Founding of the United Nations, August 1941 – October 1945" (http://web.archive.org/web/20070612221444/http://www.state.gov/r/pa/ho/pubs/fs/55407.htm). U.S. Dept. of State, Bureau of Public Affairs, Office of the Historian. October 2005. Archived from the original (http://www.state.gov/r/pa/ho/pubs/fs/55407.htm) on 2007-06-12. . Retrieved 2007-06-11.

[49] Pacific War Research Society (2006). *Japan's Longest Day*. New York: Oxford University Press. ISBN 4-7700-2887-3.

[50] Voyce, Bill (2006-08-21). "Why the Expansion of the 1990s Lasted So Long" (http://iwin.iwd.state.ia.us/iowa/ArticleReader?itemid=00003700&print=1). Iowa Workforce Information Network. . Retrieved 2007-08-16.

[51] "Many Europeans Oppose War in Iraq" (http://www.usatoday.com/news/world/2003-02-14-eu-survey.htm). *USA Today*. 2003-02-14. . Retrieved 2008-09-01.

[52] Springford, John (December 2003). "'Old' and 'New' Europeans United: Public Attitudes Towards the Iraq War and US Foreign Policy" (http://www.cer.org.uk/pdf/back_brief_springford_dec03.pdf). Centre for European Reform. . Retrieved 2008-09-01.

[53] Scheb, John M., and John M. Scheb II (2002). *An Introduction to the American Legal System*. Florence, KY: Delmar, p. 6. ISBN 0-7668-2759-3.

[54] "Brotherhood of the Burger Kings: Obama and Cameron Hail 'Stronger than Ever' Special Relationship" (http://www.dailymail.co.uk/news/article-1390800/Obama-reaffirms-special-relationship-US-historic-speech-MPs.html). Mail Online. 2011-05-26. . Retrieved 2011-08-28.

[55] Ek, Carl, and Ian F. Fergusson (2010-09-03). "Canada–U.S. Relations" (http://www.fas.org/sgp/crs/row/96-397.pdf). Congressional Research Service. . Retrieved 2011-08-28.

[56] Vaughn, Bruce (2008-08-08). "Australia: Background and U.S. Relations" (http://www.worldcat.org/title/australia-background-and-us-relations/oclc/70208969?title=&detail=&page=frame&url=http://www.fas.org/sgp/crs/row/RL33010.pdf&checksum=f2a13dd063242d8cf4b00dfda18441af&linktype=digitalObject). Congressional Research Service. . Retrieved 2011-08-28.

[57] Vaughn, Bruce (2011-05-27). "New Zealand: Background and Bilateral Relations with the United States" (http://www.fas.org/sgp/crs/row/RL32876.pdf). Congressional Research Service. . Retrieved 2011-08-28.

[58] Lum, Thomas (2011-01-03). "The Republic of the Philippines and U.S. Interests" (http://www.fas.org/sgp/crs/row/RL33233.pdf). Congressional Research Service. . Retrieved 2011-08-03.

[59] Chanlett-Avery, Emma, et al. (2011-06-08). "Japan-U.S. Relations: Issues for Congress" (http://www.fas.org/sgp/crs/row/RL33436.pdf). Congressional Research Service. . Retrieved 2011-08-28.

[60] Manyin, Mark E., Emma Chanlett-Avery, and Mary Beth Nikitin (2011-07-08). "U.S.–South Korea Relations: Issues for Congress" (http://www.fas.org/sgp/crs/row/R41481.pdf). Congressional Research Service. . Retrieved 2011-08-28.

[61] Addis, Casey L. (2011-02-14). "Israel: Background and U.S. Relations" (http://www.fas.org/sgp/crs/mideast/RL33476.pdf). Congressional Research Service. . Retrieved 2011-08-28.

[62] Shah, Anup (2009-04-13). "US and Foreign Aid Assistance" (http://www.globalissues.org/article/35/us-and-foreign-aid-assistance). GlobalIssues.org. . Retrieved 2009-10-11.

[63] "The Air Force in Facts and Figures (Armed Forces Manpower Trends, End Strength in Thousands)" (http://www.airforce-magazine.com/MagazineArchive/Magazine Documents/2009/May 2009/0509facts_fig.pdf). *Air Force Magazine*. May 2009. . Retrieved 2009-10-09.

[64] "Base Structure Report, Fiscal Year 2008 Baseline" (http://www.defense.gov/pubs/BSR_2008_Baseline.pdf). Department of Defense. . Retrieved 2009-10-09.

[65] "Active Duty Military Personnel Strengths by Regional Area and by Country (309A)" (http://siadapp.dmdc.osd.mil/personnel/MILITARY/history/hst1003.pdf). Department of Defense. 2010-03-31. . Retrieved 2010-10-07.

[66] Ikenberry, G. John (March/April 2004). "Illusions of Empire: Defining the New American Order" (http://people.cas.sc.edu/rosati/ttp.ikenberry.empirereviews.fa.march04.htm). *Foreign Affairs*. . Kreisler, Harry, and Chalmers Johnson (2004-01-29). "Conversations with History" (http://globetrotter.berkeley.edu/people4/CJohnson/cjohnson-con3.html). University of California at Berkeley. . Retrieved 2007-06-21.

[67] "The Fifteen Major Spender Countries in 2008" (http://www.sipri.org/research/armaments/milex/resultoutput/15majorspenders). Stockholm International Peace Research Institute. . Retrieved 2009-10-09.

[68] "Fiscal Year 2012 Budget Request Overview" (http://comptroller.defense.gov/defbudget/fy2012/FY2012_Budget_Request_Overview_Book.pdf). Department of Defense. February 2011. . Retrieved 2011-07-25.

[69] Hennessey, Kathleen, and Liz Sly (2010-08-02). "U.S. Combat Forces To Be out of Iraq by Aug. 31" (http://seattletimes.nwsource.com/html/nationworld/2012516655_obama03.html). Seattle Times/Tribune Washington Bureau. . Retrieved 2010-08-18.

[70] "Operation Iraqi Freedom" (http://icasualties.org/Iraq/index.aspx). Iraq Coalition Casualty Count. 2011-07-24. . Retrieved 2011-07-25.

[71] "Operation Enduring Freedom" (http://icasualties.org/OEF/index.aspx). Iraq Coalition Casualty Count. 2011-07-25. . Retrieved 2011-07-25.

[72] "Employment Situation Summary" (http://www.bls.gov/news.release/empsit.nr0.htm). U.S. Dept. of Labor. 2011-10-07. . Retrieved 2011-10-10.

[73] "Gross Domestic Product, Second Quarter 2011 (Advance Estimate)" (http://www.bea.gov/newsreleases/national/gdp/gdpnewsrelease.htm). Bureau of Economic Analysis. 2011-07-29. . Retrieved 2011-08-24. Change is based on chained 2005 dollars. Quarterly growth is expressed as an annualized rate.

[74] "Consumer Price Index: August 2011" (http://www.bls.gov/news.release/cpi.nr0.htm). Bureau of Labor Statistics. 2011-09-15. . Retrieved 2011-09-19.

[75] "Income, Poverty, and Health Insurance Coverage in the United States: 2010" (http://www.census.gov/prod/2011pubs/p60-239.pdf). U.S. Census Bureau. 2010-09-14. . Retrieved 2011-09-16.

[76] "Debt Statistics" (http://www.treasurydirect.gov/NP/BPDLogin?application=np). U.S. Dept. of the Treasury. . Retrieved 2011-09-19.

[77] "Flow of Funds Accounts of the United States" (http://www.federalreserve.gov/releases/Z1/Current/z1r-1.pdf). U.S. Federal Reserve. 2011-06-09. . Retrieved 2011-07-25.

[78] Wright, Gavin, and Jesse Czelusta, "Resource-Based Growth Past and Present", in *Natural Resources: Neither Curse Nor Destiny*, ed. Daniel Lederman and William Maloney (World Bank, 2007), p. 185. ISBN 0-8213-6545-2.

[79] "Trade Statistics" (http://greyhill.com/trade-statistics). Greyhill Advisors. . Retrieved 2011-10-06.

[80] "Top Ten Countries with which the U.S. Trades" (http://www.census.gov/foreign-trade/top/dst/current/balance.html). U.S. Census Bureau. August 2009. . Retrieved 2009-10-12.

[81] "National debt: Whom does the US owe?" (http://www.csmonitor.com/USA/Politics/DC-Decoder/2011/0204/National-debt-Whom-does-the-US-owe). CSMonitor.com. 2011-02-04. . Retrieved 2011-07-14.

[82] "New Release/Ultra Petroleum Corp.," (http://ir.nyse.com/phoenix.zhtml?c=129145&p=irol-newsArticle&ID=1036503&highlight=). NYSE Euronext. 2007-07-03. . Retrieved 2007-08-03.

[83] "Government Spending Overview" (http://www.usgovernmentspending.com/index.php). usgovernmentspending.com. . Retrieved 2009-05-09.

[84] "USA Economy in Brief" (http://web.archive.org/web/20080312123609/http://usinfo.state.gov/products/pubs/economy-in-brief/page3.html). U.S. Dept. of State, International Information Programs. Archived from the original (http://usinfo.state.gov/products/pubs/economy-in-brief/page3.html) on 2008-03-12. . Retrieved 2008-03-12.

[85] "Table 724—Number of Tax Returns, Receipts, and Net Income by Type of Business and Industry: 2005" (http://www.census.gov/compendia/statab/2009/tables/09s0724.xls). U.S. Census Bureau. . Retrieved 2009-10-12.

[86] "Table 964—Gross Domestic Product in Current and Real (2000) Dollars by Industry: 2006" (http://www.census.gov/compendia/statab/2009/tables/09s0964.xls). U.S. Census Bureau. May 2008. . Retrieved 2009-10-12.

[87] "Rank Order—Oil (Production)" (https://www.cia.gov/library/publications/the-world-factbook/rankorder/2173rank.html). *The World Factbook*. CIA. . Retrieved 2009-10-12. "Rank Order—Oil (Consumption)" (https://www.cia.gov/library/publications/the-world-factbook/rankorder/2174rank.html). *The World Factbook*. CIA. . Retrieved 2009-10-12. "Crude Oil and Total Petroleum Imports Top 15 Countries" (http://www.eia.doe.gov/pub/oil_gas/petroleum/data_publications/company_level_imports/current/import.html). U.S. Energy Information Administration. 2009-09-29. . Retrieved 2009-10-12.

[88] "Corn" (http://web.archive.org/web/20080112182404/http://www.grains.org/page.ww?section=Barley,+Corn+&+Sorghum&name=Corn). U.S. Grains Council. Archived from the original (http://www.grains.org/page.ww?section=Barley,+Corn+&+Sorghum&name=Corn) on 2008-01-12. . Retrieved 2008-03-13.

[89] "Soybean Demand Continues to Drive Production" (http://www.worldwatch.org/node/5442). Worldwatch Institute. 2007-11-06. . Retrieved 2008-03-13.

[90] "Sony, LG, Wal-Mart among Most Extendible Brands" (http://www.cheskin.com/view_news.php?id=2). Cheskin. 2005-06-06. . Retrieved 2007-06-19.

[91] Fuller, Thomas (2005-06-15). "In the East, Many EU Work Rules Don't Apply" (http://www.iht.com/articles/2005/06/14/news/europe.php). *International Herald Tribune*. . Retrieved 2007-06-28.

[92] "Doing Business in the United States (2006)" (http://www.doingbusiness.org/ExploreEconomies/?economyid=197). World Bank. . Retrieved 2007-06-28.

[93] "Total Economy Database, Summary Statistics, 1995–2010" (http://www.conference-board.org/data/economydatabase/). *The Conference Board Total Economy Database*. The Conference Board. September 2010. . Retrieved 2009-09-20.

[94] Gumbel, Peter (2004-07-11). "Escape from Tax Hell" (http://www.time.com/time/magazine/article/0,9171,662737-2,00.html). *Time*. . Retrieved 2007-06-28.

[95] DeNavas-Walt, Carmen, Bernadette D. Proctor, and Jessica Smith (August 2008). "Income, Poverty, and Health Insurance Coverage in the United States: 2007" (http://www.census.gov/prod/2008pubs/p60-235.pdf) (PDF). U.S. Census Bureau. . Retrieved 2008-11-13.

[96] Hacker, Jacob S. (2006). *The Great Risk Shift: The New Economic Insecurity and the Decline of the American Dream*. New York: Oxford University Press. ISBN 0195335341.

[97] Smeeding, T. M. (2005). "Public Policy: Economic Inequality and Poverty: The United States in Comparative Perspective". *Social Science Quarterly* **86**: 955–983. doi:10.1111/j.0038-4941.2005.00331.x.

[98] Kenworthy, L. (1999). "Do Social-Welfare Policies Reduce Poverty? A Cross-National Assessment" *Social Forces* 77(3), 1119–1139. Bradley, D., E. Huber, S. Moller, F. Nielsen, and J. D. Stephens (2003). "Determinants of Relative Poverty in Advanced Capitalist Democracies". *American Sociological Review* **68** (1): 22–51.

[99] Fishback, Price V. (May 2010). "Social Welfare Expenditures in the United States and the Nordic Countries: 1900–2003" (http://papers.nber.org/papers/w15982). *NBER Working Paper series* **15982**. .

[100] Orr, D. (November–December, 2004). "Social Security Isn't Broken: So Why the Rush to 'Fix' It?" In C. Sturr and R. Vasudevan, eds. (2007). *Current Economic Issues*. Boston: Economic Affairs Bureau.

[101] Starr, Paul (2008-02-25). "A New Deal of Their Own" (http://www.prospect.org/cs/articles?article=a_new_deal_of_their_own). *American Prospect*. . Retrieved 2008-07-24.

[102] UNICEF (2007). "Child Poverty in Perspective: An Overview of Child Well-Being in Rich Countries" (http://news.bbc.co.uk/nol/shared/bsp/hi/pdfs/13_02_07_nn_unicef.pdf). *BBC*. . Retrieved 2007-09-10.

[103] Hartman, C. (2008). "By the Numbers: Income" (http://www.demos.org/inequality/numbers.cfm#1). . Retrieved 2008-07-24.

[104] Bartels, L. M. (2008). *Unequal Democracy: The Political Economy of the New Gilded Age*. Princeton, NJ: Princeton University Press, p. 20.

[105] Henderson, David R. (1998). "The Rich—and Poor—Are Getting Richer" (http://www.hoover.org/publications/digest/3522596.html). *Hoover Digest*. . Retrieved 2007-06-19.

[106] Yellen, J. (2006). "Speech to the Center for the Study of Democracy 2006–2007 Economics of Governance Lecture University of California, Irvine" (http://www.frbsf.org/news/speeches/2006/1106.html). San Francisco: Federal Reserve Board. . Retrieved 2008-07-24.Shapiro, Isaac (2005-10-17). "New IRS Data Show Income Inequality Is Again on the Rise" (http://www.cbpp.org/cms/?fa=view&id=746). Center on Budget and Policy Priorities. . Retrieved 2007-05-16.

[107] Johnston, David Cay (2007-03-29). "Income Gap Is Widening, Data Shows" (http://www.nytimes.com/2007/03/29/business/29tax.html?ex=1332820800&en=fb472e72466c34c8&ei=5088&partner=rssnyt&emc=rss). *New York Times*. . Retrieved 2007-05-16.

[108] Saez, E. (October 2007). "Table A1: Top Fractiles Income Shares (Excluding Capital Gains) in the U.S., 1913–2005" (http://elsa.berkeley.edu/~saez/TabFig2005prel.xls). UC Berkeley. . Retrieved 2008-07-24. "Field Listing—Distribution of Family Income—Gini Index" (https://www.cia.gov/library/publications/the-world-factbook/fields/2172.html). *The World Factbook*. CIA. 2007-06-14. . Retrieved 2007-06-17.

[109] "Shares of Federal Tax Liabilities, 2004 and 2005" (http://www.cbo.gov/ftpdocs/88xx/doc8885/EffectiveTaxRates.shtml). Congressional Budget Office. . Retrieved 2008-11-02.

[110] Domhoff, G. William (December 2006). "Table 4: Percentage of Wealth Held by the Top 10% of the Adult Population in Various Western Countries" (http://sociology.ucsc.edu/whorulesamerica/power/wealth.html). *Power in America*. University of California at Santa Cruz, Sociology Dept.. . Retrieved 2006-08-21.

[111] Kennickell, Arthur B. (2006-08-02). "Table11a: Amounts (Billions of 2004 Dollars) and Shares of Net Worth and Components Distributed by Net Worth Groups, 2004" (http://www.federalreserve.gov/pubs/oss/oss2/papers/concentration.2004.5.pdf). *Currents and Undercurrents: Changes in the Distribution of Wealth, 1989–2004*. Federal Reserve Board. . Retrieved 2007-06-24.

[112] "Statistics | Inequality-adjusted HDI | Human Development Reports (HDR) | United Nations Development Programme (UNDP)" (http://hdr.undp.org/en/statistics/ihdi/). Hdr.undp.org. . Retrieved 2010-11-05.

[113] Benedetti, François (2003-12-17). "100 Years Ago, the Dream of Icarus Became Reality" (http://web.archive.org/web/20070912065254/http://www.fai.org/news_archives/fai/000295.asp). Fédération Aéronautique Internationale (FAI). Archived from the

original (http://www.fai.org/news_archives/fai/000295.asp) on 2007-09-12. . Retrieved 2007-08-15.

[114] "Research and Development (R&D) Expenditures by Source and Objective: 1970 to 2004" (https://www.census.gov/compendia/statab/2008/tables/08s0775.xls). U.S. Census Bureau. . Retrieved 2007-06-19.

[115] MacLeod, Donald (2006-03-21). "Britain Second in World Research Rankings" (http://www.guardian.co.uk/education/2006/mar/21/highereducation.uk4). *Guardian* (London). . Retrieved 2006-05-14.

[116] "Media Statistics: Televisions (per capita) by Country" (http://www.nationmaster.com/graph/med_tel_percap-media-televisions-per-capita). NationMaster. December 2003. . "Media Statistics > Personal Computers (per capita) by Country" (http://www.nationmaster.com/graph/med_per_com_percap-media-personal-computers-per-capita). NationMaster. December 2003. . "Media Statistics > Radios (per capita) by Country" (http://www.nationmaster.com/graph/med_rad_percap-media-radios-per-capita). NationMaster. December 2003. . Retrieved 2007-06-03.

[117] "Download 2007 Digital Fact Pack" (http://adage.com/digital/article?article_id=116136). *Advertising Age*. 2007-04-23. . Retrieved 2007-06-10.

[118] "ISAAA Brief 39-2008: Executive Summary—Global Status of Commercialized Biotech/GM Crops: 2008" (http://www.isaaa.org/resources/publications/briefs/39/download/isaaa-brief-39-2008.pdf). International Service for the Acquisition of Agri-Biotech Applications. p. 15. . Retrieved 2010-07-16.

[119] "Interstate FAQ (Question #3)" (http://www.fhwa.dot.gov/interstate/faq.htm#question3). Federal Highway Administration. 2006. . Retrieved 2009-03-04.

[120] "Number of Highways" (http://en.wikipedia.org/wiki/Wikispeedia). WikiSPEEDia, derived from U.S. Census Bureau. . Retrieved 2010-06-10.

[121] Newgeography.com (2011-01-22). "China Expressway System to Exceed US Interstates" (http://www.newgeography.com/content/002003-china-expressway-system-exceed-us-interstates). Newgeography.com. . Retrieved 2011-09-16.

[122] Agencies. "China overtakes US in car sales | Business | guardian.co.uk" (http://www.guardian.co.uk/business/2010/jan/08/china-us-car-sales-overtakes). Guardian. . Retrieved 2011-07-10.

[123] "Motor vehicles statistics – countries compared worldwide" (http://www.nationmaster.com/graph/tra_mot_veh-transportation-motor-vehicles). NationMaster. . Retrieved 2011-07-10.

[124] "Household, Individual, and Vehicle Characteristics" (http://www.bts.gov/publications/highlights_of_the_2001_national_household_travel_survey/html/section_01.html). *2001 National Household Travel Survey*. U.S. Dept. of Transportation, Bureau of Transportation Statistics. . Retrieved 2007-08-15.

[125] "Daily Passenger Travel" (http://www.bts.gov/publications/highlights_of_the_2001_national_household_travel_survey/html/section_02.html). *2001 National Household Travel Survey*. U.S. Dept. of Transportation, Bureau of Transportation Statistics. . Retrieved 2007-08-15.

[126] Renne, John L., and Jan S. Wells (2003). "Emerging European-Style Planning in the United States: Transit-Oriented Development (p. 2)" (http://www.policy.rutgers.edu/vtc/documents/TOD.Euro-Style_Planning-Renne-Wells.pdf) (PDF). Rutgers, The State University of New Jersey. . Retrieved 2007-06-11.

[127] "NatGeo surveys countries' transit use: guess who comes in last | Kaid Benfield's Blog | Switchboard, from NRDC" (http://switchboard.nrdc.org/blogs/kbenfield/natgeo_surveys_countries_trans.html). Switchboard.nrdc.org. 2009-05-18. . Retrieved 2011-07-10.

[128] "Intercity Passenger Rail: National Policy and Strategies Needed to Maximize Public Benefits from Federal Expenditures" (http://www.gao.gov/products/GAO-07-15). U.S. Government Accountability Office. 2006-11-13. . Retrieved 2007-06-20.

[129] McGill, Tracy (2011-01-01). "3 Reasons Light Rail Is an Efficient Transportation Option for U.S. Cities" (http://www.metaefficient.com/trains/master-2.html). MetaEfficient. . Retrieved 2011-08-29.

[130] "Bicycling to Work" (http://trafficsafety.org/safety/sharing/bike/bike-facts/bicycling-to-work). Trafficsafety.org. . Retrieved 2011-07-10.

[131] "Scheduled Passengers Carried (2008 data)" (http://www.iata.org/ps/publications/wats-passenger-carried.htm). International Air Transport Association (IATA). . Retrieved 2009-06-27.

[132] "Passenger Traffic 2006 Final" (http://www.airports.org/cda/aci_common/display/main/aci_content07_c.jsp?zn=aci&cp=1-5-54-55_666_2__). Airports Council International. 2007-07-18. . Retrieved 2007-08-15.

[133] "Diagram 1: Energy Flow, 2007" (http://www.eia.doe.gov/emeu/aer/pdf/pages/sec1_3.pdf). *EIA Annual Energy Review 2007*. U.S. Dept. of Energy, Energy Information Administration. . Retrieved 2008-06-25.

[134] "Rank Order—Oil (Consumption)" (https://www.cia.gov/library/publications/the-world-factbook/rankorder/2174rank.html). *The World Factbook*. CIA. 2007-09-06. . Retrieved 2007-09-14.

[135] "Atomic Renaissance" (http://www.economist.com/science/displaystory.cfm?story_id=9762843). *Economist*. 2007-09-06. . Retrieved 2007-09-06.

[136] "BP Statistical Review of World Energy" (http://www.bp.com/liveassets/bp_internet/globalbp/globalbp_uk_english/reports_and_publications/statistical_energy_review_2007/STAGING/local_assets/downloads/spreadsheets/statistical_review_full_report_workbook_2007.xls) (XLS). British Petroleum. June June 2007. . Retrieved 2010-02-22.

[137] "The Times Higher Education University Rankings 2010–2011" (http://www.timeshighereducation.co.uk/world-university-rankings/2010-2011/top-200.html). Times Higher Education. . Retrieved 2011-08-01. "World University Rankings 2010" (http://www.topuniversities.com/university-rankings/world-university-rankings/2010). Quacquarelli Symonds. . Retrieved 2011-08-01.

[138] "Ages for Compulsory School Attendance..." (http://nces.ed.gov/programs/digest/d02/dt150.asp). U.S. Dept. of Education, National Center for Education Statistics. . Retrieved 2007-06-10.

[139] "Statistics About Non-Public Education in the United States" (http://www2.ed.gov/about/offices/list/oii/nonpublic/statistics.html). U.S. Dept. of Education, Office of Non-Public Education. . Retrieved 2007-06-05.

[140] "QS World University Rankings" (http://www.topuniversities.com/university-rankings/world-university-rankings/2010). Topuniversities. . Retrieved 2011-07-10.

[141] "Top 200 – The Times Higher Education World University Rankings 2010–2011" (http://www.timeshighereducation.co.uk/world-university-rankings/2010-2011/top-200.html). Timeshighereducation.co.uk. . Retrieved 2011-07-10.

[142] "Educational Attainment in the United States: 2003" (http://www.census.gov/prod/2004pubs/p20-550.pdf). U.S. Census Bureau. . Retrieved 2006-08-01.

[143] For more detail on U.S. literacy, see A First Look at the Literacy of America's Adults in the 21st century (http://nces.ed.gov/NAAL/PDF/2006470.PDF), U.S. Department of Education (2003).

[144] "Human Development Indicators" (http://web.archive.org/web/20070620235428/http://hdr.undp.org/reports/global/2005/pdf/HDR05_HDI.pdf). United Nations Development Programme, Human Development Reports. 2005. Archived from the original (http://hdr.undp.org/reports/global/2005/pdf/HDR05_HDI.pdf) on 2007-06-20. . Retrieved 2008-01-14.

[145] Eberstadt, Nicholas, and Hans Groth (2007-04-19). "Healthy Old Europe" (http://www.iht.com/articles/2007/04/19/opinion/edeber.php). *International Herald Tribune*. . Retrieved 2007-06-19.

[146] MacAskill, Ewen (2007-08-13). "US Tumbles Down the World Ratings List for Life Expectancy" (http://www.guardian.co.uk/world/2007/aug/13/usa.ewenmacaskill). *Guardian* (London). . Retrieved 2007-08-15.

[147] "Rank Order—Infant Mortality Rate" (https://www.cia.gov/library/publications/the-world-factbook/rankorder/2091rank.html). *The World Factbook*. CIA. 2007-06-14. . Retrieved 2007-06-19.

[148] "Prevalence of Overweight and Obesity Among Adults: United States, 2003–2004" (http://www.cdc.gov/nchs/products/pubs/pubd/hestats/overweight/overwght_adult_03.htm). Centers for Disease Control and Prevention, National Center for Health Statistics. . Retrieved 2007-06-05.

[149] Schlosser, Eric (2002). *Fast Food Nation*. New York: Perennial. p. 240. ISBN 0060938455.

[150] "Fast Food, Central Nervous System Insulin Resistance, and Obesity" (http://atvb.ahajournals.org/cgi/content/full/25/12/2451#R3-101329). *Arteriosclerosis, Thrombosis, and Vascular Biology*. American Heart Association. 2005. . Retrieved 2007-06-17.

[151] "2007 Facts & Figures" (http://www.texmedctr.tmc.edu/root/en/GetToKnow/FactsandFigures/Facts+and+Figures.htm). Texas Medical Center. . Retrieved 2008-11-07.

[152] *OECD Health Data 2000: A Comparative Analysis of 29 Countries* [CD-ROM] (OECD: Paris, 2000). See also "The U.S. Healthcare System: The Best in the World or Just the Most Expensive?" (http://classic-web.archive.org/web/20061118234952/http://dll.umaine.edu/ble/U.S.+HCweb.pdf). University of Maine. 2001. . Retrieved 2006-11-29.

[153] "Health, United States, 2006" (http://www.cdc.gov/nchs/data/hus/hus06.pdf). Centers for Disease Control, National Center for Health Statistics. . Retrieved 2006-11-24.

[154] "Poverty Remains Higher, and Median Income for Non-Elderly Is Lower, Than When Recession Hit Bottom: Poor Performance Unprecedented for Four-Year Recovery Period" (http://www.cbpp.org/cms/?fa=view&id=629). Center for Budget and Policy Priorities. 2006-09-01. . Retrieved 2007-06-24.

[155] Abelson, Reed (2008-06-10). "Ranks of Underinsured Are Rising, Study Finds" (http://www.nytimes.com/2008/06/10/health/policy/10health.html). *New York Times*. . Retrieved 2008-10-25. Blewett, Lynn A. et al., LA (Dec 2006). "How Much Health Insurance Is Enough? Revisiting the Concept of Underinsurance". *Medical Care Research and Review* **63** (6): 663–700. doi:10.1177/1077558706293634. ISSN 1077-5587. PMID 17099121.

[156] Park, Madison (2009-09-18). "45,000 American Deaths Associated with Lack of Insurance" (http://www.cnn.com/2009/HEALTH/09/18/deaths.health.insurance/index.html?eref=rss_latest). CNN. . Retrieved 2009-10-01.

[157] Fahrenthold, David A. (2006-04-05). "Mass. Bill Requires Health Coverage" (http://www.washingtonpost.com/wp-dyn/content/article/2006/04/04/AR2006040401937.html). *Washington Post*. . Retrieved 2007-06-19.

[158] "Eighth United Nations Survey of Crime Trends and Operations of Criminal Justice Systems (2001–2002)" (http://www.unodc.org/pdf/crime/eighthsurvey/8sv.pdf). United Nations Office on Drugs and Crime (UNODC). 2005-03-31. . Retrieved 2008-05-18.

[159] "Murder" (http://www2.fbi.gov/ucr/cius2009/offenses/violent_crime/murder_homicide.html). *Crime in the United States 2009*. FBI. September 2010. . Retrieved 2011-07-23.

[160] "New Incarceration Figures: Thirty-Three Consecutive Years of Growth" (http://www.sentencingproject.org/doc/publications/inc_newfigures.pdf). Sentencing Project. December 2006. . Retrieved 2007-06-10.

[161] Walmsley, Roy (2005). "World Prison Population List" (http://web.archive.org/web/20070628215935/http://www.kcl.ac.uk/depsta/rel/icps/world-prison-population-list-2005.pdf). King's College London, International Centre for Prison Studies. Archived from the original (http://www.kcl.ac.uk/depsta/rel/icps/world-prison-population-list-2005.pdf) on 2007-06-28. . Retrieved 2007-10-19. For the latest data, see "Prison Brief for United States of America" (http://web.archive.org/web/20070804061423/http://www.kcl.ac.uk/depsta/rel/icps/worldbrief/north_america_records.php?code=190). King's College London, International Centre for Prison Studies. 2006-06-21. Archived from the original (http://www.kcl.ac.uk/depsta/rel/icps/worldbrief/north_america_records.php?code=190) on 2007-08-04. . Retrieved 2007-10-19. For other estimates of the incarceration rate in China and North Korea see Adams, Cecil (2004-02-06). "Does the United States Lead the World in Prison Population?" (http://www.straightdope.com/columns/read/2494/

does-the-united-states-lead-the-world-in-prison-population). The Straight Dope. . Retrieved 2007-10-11.

[162] "Pew Report Finds More than One in 100 Adults are Behind Bars" (http://www.pewcenteronthestates.org/news_room_detail.aspx?id=35912). Pew Center on the States. 2008-02-28. . Retrieved 2008-03-02.

[163] "Incarceration Rate, 1980–2005" (http://web.archive.org/web/20070611192107/http://www.ojp.usdoj.gov/bjs/glance/tables/incrttab.htm). U.S. Dept. of Justice, Bureau of Justice Statistics. 2006. Archived from the original (http://www.ojp.usdoj.gov/bjs/glance/tables/incrttab.htm) on 2007-06-11. . Retrieved 2007-06-10.

[164] "World Prison Population List (8th edition)" (http://www.prisonstudies.org/info/downloads/wppl-8th_41.pdf). King's College London, International Centre for Prison Studies. 2009. . Retrieved 2011-07-23.

[165] "The Impact of the War on Drugs on U.S. Incarceration" (http://www.hrw.org/legacy/reports/2000/usa/Rcedrg00-05.htm). Human Rights Watch. May 2000. . Retrieved 2007-06-10.

[166] "Executions in the United States in 2007" (http://www.deathpenaltyinfo.org/executions-united-states-2007). Death Penalty Information Center. . Retrieved 2007-06-15.

[167] "Executions Around the World" (http://www.deathpenaltyinfo.org/death-penalty-international-perspective#interexec). Death Penalty Information Center. 2010. . Retrieved 2011-07-23.

[168] "Quinn Signs Death Penalty Ban, Commutes 15 Death Row Sentences to Life" (http://newsblogs.chicagotribune.com/clout_st/2011/03/quinn-signs-death-penalty-ban-commutes-15-death-row-sentences-to-life.html). *Chicago Tribune*. 2011-03-09. . Retrieved 2011-03-09.

[169] "2010 Census Data" (http://2010.census.gov/2010census/data/). U.S. Census Bureau. . Retrieved 2011-03-29.

[170] Camarota, Steven A., and Karen Jensenius (July 2008). "Homeward Bound: Recent Immigration Enforcement and the Decline in the Illegal Alien Population" (http://www.cis.org/articles/2008/back808.pdf). Center for Immigration Studies. . Retrieved 2008-08-06.

[171] "Statistical Abstract of the United States" (http://www.census.gov/prod/2005pubs/06statab/pop.pdf) (PDF). United States Census Bureau. .

[172] "Executive Summary: A Population Perspective of the United States" (http://web.archive.org/web/20070604165856/http://www.prcdc.org/summaries/uspopperspec/uspopperspec.html). Population Resource Center. May 2000. Archived from the original (http://www.prcdc.org/summaries/uspopperspec/uspopperspec.html) on 2007-06-04. . Retrieved 2007-12-20.

[173] "Rank Order—Birth Rate" (https://www.cia.gov/library/publications/the-world-factbook/rankorder/2054rank.html). *The World Factbook*. CIA. 2009. . Retrieved 2009-10-11.

[174] "U.S. Legal Permanent Residents: 2010" (http://www.dhs.gov/xlibrary/assets/statistics/publications/lpr_fr_2010.pdf). Office of Immigration Statistics *Annual Flow Report.*

[175] "Persons Obtaining Legal Permanent Resident Status by Region and Country of Birth: Fiscal Years 1998 to 2007 (Table 3)" (http://www.dhs.gov/files/statistics/publications/LPR07.shtm). U.S. Dept. of Homeland Security. . Retrieved 2008-09-06.

[176] "Ancestry 2000" (http://www.census.gov/prod/2004pubs/c2kbr-35.pdf). U.S.Census Bureau. June 2004. . Retrieved 2007-06-13.

[177] Humes, Karen R., Nicholas A. Jones, and Roberto R. Ramirez (March 2011). "Overview of Race and Hispanic Origin: 2010" (http://www.census.gov/prod/cen2010/briefs/c2010br-02.pdf). U.S. Census Bureau. . Retrieved 2011-03-29.

[178] "B03001. Hispanic or Latino Origin by Specific Origin" (http://factfinder.census.gov/servlet/DTTable?_bm=y&-ds_name=ACS_2007_1YR_G00_&-CONTEXT=dt&-mt_name=ACS_2007_1YR_G2000_B03001&-redoLog=true&-geo_id=01000US&-geo_id=04000US48&-format=&-_lang=en&-SubjectID=15233304). *2007 American Community Survey*. U.S. Census Bureau. . Retrieved 2008-09-26.

[179] "Tables 41 and 42—Native and Foreign-Born Populations" (http://www.census.gov/prod/2008pubs/09statab/pop.pdf). *Statistical Abstract of the United States 2009*. U.S. Census Bureau. . Retrieved 2009-10-11.

[180] US Census Bureau: "U.S. Census Bureau Delivers Final State 2010 Census Population Totals for Legislative Redistricting" (http://2010.census.gov/news/releases/operations/cb11-cn123.html) see custom table, 2nd worksheet

[181] "An Older and More Diverse Nation by Midcentury" (http://web.archive.org/web/20080822044429/http://www.census.gov/Press-Release/www/releases/archives/population/012496.html). U.S. Census Bureau. 2008-08-14. Archived from the original (http://www.census.gov/Press-Release/www/releases/archives/population/012496.html) on 2008-08-22. . Retrieved 2008-09-06.

[182] "United States—Urban/Rural and Inside/Outside Metropolitan Area (GCT-P1. Population, Housing Units, Area, and Density: 2000)" (http://factfinder.census.gov/servlet/GCTTable?_bm=y&-state=gct&-ds_name=DEC_2000_SF1_U&-_box_head_nbr=GCT-P1&-mt_name=&-_caller=geoselect&-geo_id=&-format=US-1&-_lang=en). U.S. Census Bureau. 2000-04-01. . Retrieved 2008-09-23.

[183] "Table 1: Annual Estimates of the Resident Population for Incorporated Places Over 100,000, Ranked by July 1, 2008 Population: April 1, 2000 to July 1, 2008" (http://www.webcitation.org/5lpvuJk99) (PDF). *2008 Population Estimates*. U.S. Census Bureau, Population Division. 2009-07-01. Archived from the original (http://hawaii.gov/dbedt/info/census/popestimate/copy_of_2008-subcounty-population-hawaii/SUB_EST2008_01.pdf) on 2009-12-07. . Retrieved 2009-10-11.

[184] "Table 5. Estimates of Population Change for Metropolitan Statistical Areas and Rankings: July 1, 2007 to July 1, 2008" (http://www.webcitation.org/5lpvuJzkG) (PDF). *2008 Population Estimates*. U.S. Census Bureau. 2009-03-19. Archived from the original (http://hawaii.gov/dbedt/info/census/popestimate/2008_MSA_Hawaii/CBSA_EST2008_05.pdf) on 2009-12-07. . Retrieved 2009-10-11.

[185] "Raleigh and Austin are Fastest-Growing Metro Areas" (http://www.census.gov/newsroom/releases/archives/population/cb09-45.html). U.S. Census Bureau. 2009-03-19. . Retrieved 2009-10-11.

[186] "2010 Census National Summary File of Redistricting Data" (http://factfinder2.census.gov/faces/tableservices/jsf/pages/productview.xhtml?pid=DEC_10_NSRD_GCTPL2.US24PR&prodType=table). U.S. Census Bureau. . Retrieved 2011-05-05.

[187] "Figure A–3. Census Regions, Census Divisions, and Their Constituent States" (http://www.census.gov/geo/www/tiger/glossry2.pdf). U.S. Census Bureau. . Retrieved 2007-06-17.

[188] "Table 53—Languages Spoken at Home by Language: 2007" (http://www.census.gov/compendia/statab/2010/tables/10s0053.pdf). *Statistical Abstract of the United States 2010*. U.S. Census Bureau. . Retrieved 2009-09-21.

[189] "Foreign Language Enrollments in United States Institutions of Higher Learning" (http://www.adfl.org/resources/enrollments.pdf). MLA. fall 2002. . Retrieved 2006-10-16.

[190] "The Constitution of the State of Hawaii, Article XV, Section 4" (http://www.hawaii.gov/lrb/con/conart15.html). Hawaii Legislative Reference Bureau. 1978-11-07. . Retrieved 2007-06-19.

[191] Dicker, Susan J. (2003). *Languages in America: A Pluralist View*. Clevedon, UK: Multilingual Matters. pp. 216, 220–25. ISBN 1853596515.

[192] "California Code of Civil Procedure, Section 412.20(6)" (http://www.leginfo.ca.gov/cgi-bin/displaycode?section=ccp&group=00001-01000&file=412.10-412.30). Legislative Counsel, State of California. . Retrieved 2007-12-17. "California Judicial Council Forms" (http://www.courtinfo.ca.gov/forms/allforms.htm). Judicial Council, State of California. . Retrieved 2007-12-17.

[193] "Among Wealthy Nations...U.S. Stands Alone in its Embrace of Religion" (http://pewglobal.org/reports/display.php?ReportID=167). *Pew Global Attitudes Project*. Pew Research Center. 2002-12-19. . Retrieved 2008-10-23.

[194] "Religious Composition of the U.S." (http://religions.pewforum.org/pdf/affiliations-all-traditions.pdf). *U.S. Religious Landscape Survey*. Pew Forum on Religion & Public Life. 2007. . Retrieved 2008-10-23.

[195] Kosmin, Barry A., Egon Mayer, and Ariela Kaysar (2001-12-19). "American Religious Identification Survey 2001" (http://www.gc.cuny.edu/CUNY_GC/media/CUNY-Graduate-Center/PDF/ARIS/ARIS-PDF-version.pdf?ext=.pdf). CUNY Graduate Center. . Retrieved 2011-09-16.

[196] Green, John C. "The American Religious Landscape and Political Attitudes: A Baseline for 2004" (http://www.webcitation.org/5lpvuMoCD). University of Akron Ray C. Bliss Institute of Applied Politics. Archived from the original (http://www.uakron.edu/bliss/docs/Religious_Landscape_2004.pdf) on 2009-12-07. . Retrieved 2007-06-18.

[197] "Table 55—Marital Status of the Population by Sex, Race, and Hispanic Origin: 1990 to 2007" (http://www.census.gov/prod/2008pubs/09statab/pop.pdf). *Statistical Abstract of the United States 2009*. U.S. Census Bureau. . Retrieved 2009-10-11.

[198] "Women's Advances in Education" (http://web.archive.org/web/20070609151527/http://www.iserp.columbia.edu/news/articles/female_college.html). Columbia University, Institute for Social and Economic Research and Policy. 2006. Archived from the original (http://www.iserp.columbia.edu/news/articles/female_college.html) on 2007-06-09. . Retrieved 2007-06-06.

[199] "For First Time, Majority of Americans Favor Legal Gay Marriage" (http://www.gallup.com/poll/147662/First-Time-Majority-Americans-Favor-Legal-Gay-Marriage.aspx). Gallup.com. . Retrieved 2011-07-10.

[200] "Teenage birth rate statistics – countries compared – NationMaster People" (http://www.nationmaster.com/graph/peo_tee_bir_rat-people-teenage-birth-rate). Nationmaster.com. . Retrieved 2011-07-10.

[201] Strauss, Lilo T., et al. (2006-11-24). "Abortion Surveillance—United States, 2003" (http://www.cdc.gov/mmwr/preview/mmwrhtml/ss5511a1.htm). *MMWR*. Centers for Disease Control, National Center for Chronic Disease Prevention and Health Promotion, Division of Reproductive Health. . Retrieved 2007-06-17.

[202] Thompson, William, and Joseph Hickey (2005). *Society in Focus*. Boston: Pearson. ISBN 0-205-41365-X.

[203] Fiorina, Morris P., and Paul E. Peterson (2000). *The New American Democracy*. London: Longman, p. 97. ISBN 0-321-07058-5.

[204] Holloway, Joseph E. (2005). *Africanisms in American Culture*, 2d ed. Bloomington: Indiana University Press, pp. 18–38. ISBN 0-253-34479-4. Johnson, Fern L. (1999). *Speaking Culturally: Language Diversity in the United States*. Thousand Oaks, Calif., London, and New Delhi: Sage, p. 116. ISBN 0-8039-5912-5.

[205] "Individualism" (http://www.clearlycultural.com/geert-hofstede-cultural-dimensions/individualism/). Clearly Cultural. . Retrieved 2009-02-28.

[206] "A Family Affair: Intergenerational Social Mobility across OECD Countries" (http://www.oecd.org/dataoecd/2/7/45002641.pdf). *Economic Policy Reforms: Going for Growth* (OECD). 2010. . Retrieved 2010-09-20. Blanden, Jo, Paul Gregg, and Stephen Malchin (April 2005). "Intergenerational Mobility in Europe and North America" (http://web.archive.org/web/20060623094610/http://www.suttontrust.com/reports/IntergenerationalMobility.pdf). Centre for Economic Performance. Archived from the original (http://www.suttontrust.com/reports/IntergenerationalMobility.pdf) on June 23, 2006. . Retrieved 2006-08-21.

[207] Gutfield, Amon (2002). *American Exceptionalism: The Effects of Plenty on the American Experience*. Brighton and Portland: Sussex Academic Press. p. 65. ISBN 1903900085.

[208] Zweig, Michael (2004). *What's Class Got To Do With It, American Society in the Twenty-First Century*. Ithaca, NY: Cornell University Press. ISBN 0801488990. "Effects of Social Class and Interactive Setting on Maternal Speech" (http://eric.ed.gov/ERICWebPortal/Home.portal?_nfpb=true&_pageLabel=RecordDetails&ERICExtSearch_SearchValue_0=ED309843&ERICExtSearch_SearchType_0=eric_accno&objectId=0900000b800472a5). Education Resource Information Center. . Retrieved 2007-01-27.

[209] Ehrenreich, Barbara (1989). *Fear of Falling, The Inner Life of the Middle Class*. New York: HarperCollins. ISBN 0060973331.

[210] Eichar, Douglas (1989). *Occupation and Class Consciousness in America*. Westport, CT: Greenwood Press. ISBN 0313261113.

[211] O'Keefe, Kevin (2005). *The Average American*. New York: PublicAffairs. ISBN 158648270X.

[212] *Village Voice*: 100 Best Films of the 20th century (2001) (http://www.filmsite.org/villvoice.html). Filmsite.org; *Sight and Sound* Top Ten Poll 2002 (http://www.bfi.org.uk/sightandsound/topten/poll/critics-long.html). BFI. Retrieved on 2007-06-19.

[213] "World Culture Report 2000 Calls for Preservation of Intangible Cultural Heritage" (http://www.unesco.org/bpi/eng/unescopress/2000/00-120e.shtml). UNESCO. 2000-11-17. . Retrieved 2007-09-14. "Summary: Does Globalization Thwart Cultural Diversity?" (http://web.archive.org/web/20071017111347/http://worldbank.org/economicpolicy/globalization/thwart.html). World Bank Group. Archived from the original (http://www1.worldbank.org/economicpolicy/globalization/thwart.html) on 2007-10-17. . Retrieved 2007-09-14.

[214] "Media Statistics > Television Viewing by Country" (http://www.nationmaster.com/graph/med_tel_vie-media-television-viewing). NationMaster. . Retrieved 2007-06-03.

[215] "Broadband and Media Consumption" (http://www.emarketer.com/Article.aspx?R=1005003). eMarketer. 2007-06-07. . Retrieved 2007-06-10.

[216] "TV Fans Spill into Web Sites" (http://www.emarketer.com/Article.aspx?R=1004830). eMarketer. 2007-06-07. . Retrieved 2007-06-10.

[217] "Top Sites in United States" (http://www.alexa.com/topsites/countries/US). Alexa. 2010. . Retrieved 2010-03-27.

[218] Biddle, Julian (2001). *What Was Hot!: Five Decades of Pop Culture in America*. New York: Citadel, p. ix. ISBN 0-8065-2311-5.

[219] Bloom, Harold. 1999. *Emily Dickinson*. Broomall, PA: Chelsea House Publishers. p. 9. ISBN 0-7910-5106-4.

[220] Buell, Lawrence (Spring/Summer 2008). "The Unkillable Dream of the Great American Novel: *Moby-Dick* as Test Case". *American Literary History* **20** (1–2): 132–155. doi:10.1093/alh/ajn005. ISSN 0896-7148.

[221] Quinn, Edward (2006). *A Dictionary of Literary and Thematic Terms*. Infobase, p. 361. ISBN 0-8160-6243-9. Seed, David (2009). *A Companion to Twentieth-Century United States Fiction*. Chichester, West Sussex: John Wiley and Sons, p. 76. ISBN 1-4051-4691-5. Meyers, Jeffrey (1999). *Hemingway: A Biography*. New York: Da Capo, p. 139. ISBN 0-306-80890-0.

[222] Brown, Milton W. (1988 1963). *The Story of the Armory Show*. New York: Abbeville. ISBN 0-89659-795-4.

[223] Daniels, Les (1998). *Superman: The Complete History* (1st ed.). Titan Books. p. 11. ISBN 1-85286-988-7.

[224] Klapthor, James N. (2003-08-23). "What, When, and Where Americans Eat in 2003" (http://www.ift.org/cms/?pid=1000496). Institute of Food Technologists. . Retrieved 2007-06-19.

[225] Smith, Andrew F. (2004). *The Oxford Encyclopedia of Food and Drink in America*. New York: Oxford University Press, pp. 131–32. ISBN 0-19-515437-1. Levenstein, Harvey (2003). *Revolution at the Table: The Transformation of the American Diet*. Berkeley, Los Angeles, and London: University of California Press, pp. 154–55. ISBN 0-520-23439-1.

[226] Boslaugh, Sarah (2010). "Obesity Epidemic", in *Culture Wars: An Encyclopedia of Issues, Viewpoints, and Voices*, ed. Roger Chapman. Armonk, N.Y.: M. E. Sharpe, pp. 413–14. ISBN 978-0-7656-1761-3.

[227] "Fast Food, Central Nervous System Insulin Resistance, and Obesity" (http://atvb.ahajournals.org/cgi/content/full/25/12/2451#R3-101329). *Arteriosclerosis, Thrombosis, and Vascular Biology*. American Heart Association. 2005. . Retrieved 2007-06-09. "Let's Eat Out: Americans Weigh Taste, Convenience, and Nutrition" (http://www.ers.usda.gov/publications/eib19/eib19_reportsummary.pdf). U.S. Dept. of Agriculture. . Retrieved 2007-06-09.

[228] Krane, David K. (2002-10-30). "Professional Football Widens Its Lead Over Baseball as Nation's Favorite Sport" (http://www.harrisinteractive.com/Insights/HarrisVault8482.aspx?PID=337). Harris Interactive. . Retrieved 2007-09-14. Maccambridge, Michael (2004). *America's Game: The Epic Story of How Pro Football Captured a Nation*. New York: Random House. ISBN 0-375-50454-0.

[229] "All-Time Medal Standings, 1896–2004" (http://www.infoplease.com/ipsa/A0115108.html). Information Please. . Retrieved 2007-06-14. "Distribution of Medals—2008 Summer Games" (http://www.factmonster.com/sports/olympics/2008/distribution-medals-summer-games.html). Fact Monster. . Retrieved 2008-09-02.

[230] "All-Time Medal Standings, 1924–2006" (http://www.infoplease.com/ipsa/A0115207.html). Information Please. . Retrieved 2007-06-14. "Olympic Medals" (http://www.vancouver2010.com/olympic-medals/). Vancouver Organizing Committee. . Retrieved 2010-03-02. Norway is first.

[231] "Appendix G: Weights and Measures" (https://www.cia.gov/library/publications/the-world-factbook/appendix/appendix-g.html). *The World Factbook*. CIA. . Retrieved 2010-04-01.

External links

Government

- Official U.S. Government Web Portal (http://www.usa.gov/) Gateway to government sites
- House (http://www.house.gov/) Official site of the United States House of Representatives
- Senate (http://www.senate.gov/) Official site of the United States Senate
- White House (http://www.whitehouse.gov/) Official site of the President of the United States
- Supreme Court (http://www.supremecourt.gov/) Official site of the Supreme Court of the United States

Overviews and Data

- United States (https://www.cia.gov/library/publications/the-world-factbook/geos/us.html) entry at *The World Factbook*
- InfoUSA (http://infousa.state.gov/) Portal to U.S. Information Agency resources
- Library of Congress (http://www.loc.gov/index.html) Official site of the U.S. Library of Congress

- The 50 States of the U.S.A. (http://www.teacheroz.com/states.htm) Collected informational links for each state
- United States travel guide from Wikitravel
- United States (http://www.britannica.com/EBchecked/topic/616563/United-States) *Encyclopædia Britannica* entry
- United States (http://www.dmoz.org/Regional/North_America/United_States/) at the Open Directory Project
- U.S. Census Housing and Economic Statistics (http://www.census.gov/hhes/www/) Wide-ranging data from the U.S. Census Bureau
- U.S. Citizenship and Immigration Services (http://www.uscis.gov/portal/site/uscis) Official government site
- State Fact Sheets (http://www.ers.usda.gov/statefacts/) Population, employment, income, and farm data from the U.S. Economic Research Service
- State Energy Profiles (http://tonto.eia.doe.gov/state/) Economic, environmental, and energy data for each state from the U.S. Energy Information Administration
- Category:USA (http://wiki.creativecommons.org/Category:USA) in the Creative Commons. Websites and projects for free usage under a Creative Commons license.

History

- Historical Documents (http://www.nationalcenter.org/HistoricalDocuments.html) Collected by the National Center for Public Policy Research
- U.S. National Mottos: History and Constitutionality (http://www.religioustolerance.org/nat_mott.htm) Analysis by the Ontario Consultants on Religious Tolerance
- USA (http://www.historicalstatistics.org/index2.html) Collected links to historical data

Maps

- National Atlas of the United States (http://nationalatlas.gov/) Official maps from the U.S. Department of the Interior
- United States (http://www.wikimapia.org/#y=41771312&x=-99492187&z=4&l=0&m=a) Satellite view at WikiMapia (not affiliated with Wikipedia/Wikimedia Foundation)
- Wikimedia Atlas of the United States

kbd:Америкэ Штат Зэгуэт gag:Amerika Birleşik Devletläri koi:Америкаись Ӧтлаасьӧм Штаттэз ltg:Amerikys Sašķierstuos Vaļsteibys mrj:Америкын Ушымы Штатвлäжӹ pfl:Verainischde Schdaade vun Ameriga rue:Споєны Штаты Америцькы

United States Census Bureau

Bureau of the Census	
Seal	
USCENSUSBUREAU	
Logo	
Agency overview	
Formed	July 1, 1903
Preceding agency	Temporary census offices
Headquarters	Suitland, Maryland
Employees	5,593 (2006)
Annual budget	US$3.1 billion (2009) US$7.2 billion (est. 2010) US$1.3 billion (est. 2011)
Agency executive	Director, Robert Groves
Parent agency	Economics and Statistics Administration
Website	
www.census.gov [1]	

The **United States Census Bureau** (officially the **Bureau of the Census**, as defined in Title 13 U.S.C. § 11 [2]) is the government agency that is responsible for the United States Census. It also gathers other national demographic and economic data. As part of the United States Department of Commerce, the Census Bureau serves as a leading source of data about America's people and economy.[3]

The most visible role of the Census Bureau is to perform the official decennial (every 10 years) count of people living in the U.S. The most important result is the reallocation of the number of seats each state is allowed in the House of Representatives,[3] but the results also affect a range of government programs received by each state. The agency director is a political appointee selected by the President of the United States.

Legal mandate

The Constitution of the United States (Article I, section II) directs that the population be enumerated at least once every ten years and the resulting counts used to set the number of members from each state in the House of Representatives and, by extension, in the Electoral College. The Census Bureau now conducts a full population count every 10 years in years ending with a 0 (zero) and uses the term "decennial" to describe the operation. Between censuses, the Census Bureau makes population estimates and projections.[4] In addition, Census data directly affects how more than $400 billion per year in federal and state funding is allocated to communities for neighborhood improvements, public health, education, transportation and much more. The Census Bureau is mandated with fulfilling these obligations: the collecting of statistics about the nation, its people, and economy. The Census Bureau's legal authority is codified in Title 13 of the United States Code. In addition, the Census Bureau also conducts surveys on behalf of various federal government and local government agencies on topics such as employment, crime, health, consumer expenditures, and housing. Within the bureau, these are known as "demographic surveys" and are conducted perpetually between and during decennial (10-year) population counts. The Census Bureau also conducts economic surveys of manufacturing, retail, service, and other establishments and of domestic governments.

From 1790 to 1840, the census was taken by marshals of the judicial districts.[5] The Census Act of 1840 established a central office[6] which became known as the Census Office. Several acts followed that revised and authorized new censuses, typically at the 10-year intervals. In 1902 the temporary Census Office was moved under the Department of Interior, and in 1903 it was renamed the Census Bureau under the new Department of Commerce and Labor. The department was intended to consolidate overlapping statistical agencies, but Census Bureau officials were hindered by their subordinate role in the department.[7] An act in 1920 changed the date and authorized manufacturing censuses every 2 years and agriculture censuses every 10 years.[8] In 1929, a bill was passed mandating that the House of Representatives be reapportioned based on the results of the 1930 Census.[8] In 1954, various acts were codified into Title 13 of the US Code.[9]

By law the Census Bureau must count everyone and submit state population totals to the U.S. President by December 31, 2010. States within the Union receive the results in the spring of the following year.

Uses of census data

Many federal, state, local and tribal governments use census data to:

- Decide the location of new housing and public facilities,
- Examine the demographic characteristics of communities, states, and the USA,
- Plan transportation systems and roadways,
- Determine quotas and creation of police and fire precincts, and
- Create localized areas for elections, schools, utilities, etc.

Businesses

Business also has many uses for census data, as listed;

- Forecast future product demand,
- Determine site locations for expansion/new business,
- Determine future need for nursing homes, day care centers, hospitals, etc.,
- Clarify if they are employing a representative workforce.[3]

Data stewardship

The United States Census Bureau is committed to confidentiality and guarantees non-disclosure of any addresses or personal information related to individuals or establishments. Title 13 of the U.S. Code establishes penalties for the disclosure of this information. All Census employees must sign a sworn affidavit of non-disclosure prior to employment.

The Bureau cannot share responses, addresses or personal information with anyone including United States or foreign government and law enforcement agencies such as the IRS or the FBI or Interpol. "Providing quality data, for public good—while respecting individual privacy and, at the same time, protecting confidentiality—is the Census Bureau's core responsibility", "Keeping the public's trust is critical to the Census's ability to carry out the mission as the leading source of quality data about the Nation's people and economy."[10] Only after 72 years does the information collected become available to other agencies or the general public.

In 1918, the Census Bureau released individual information regarding several hundred young men to the Justice Department and Selective Service system for the purpose of prosecutions for draft evasion.[11] [12] During World War II, the United States Census Bureau assisted the government's Japanese American internment efforts by providing confidential neighborhood information on Japanese-Americans. The Bureau's role was denied for decades but was finally proven in 2007.[13] [14]

Organizational structure

Since 1903, the official census-taking agency of the United States government has been the Bureau of the Census. The Census Bureau is headed by a Director, assisted by a Deputy Director and an Executive Staff composed of the associate directors. In April 2009, President Barack Obama nominated Robert M. Groves to head the Census Bureau.

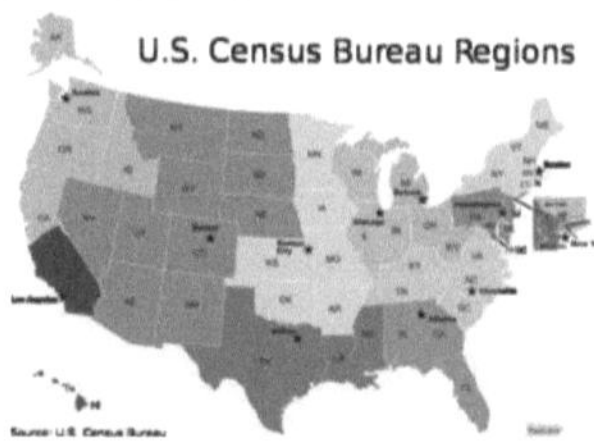

US Census Bureau Regions, centered on and divided by its twelve regional offices.

Census headquarters in Suitland, Maryland.

The Census Bureau has had headquarters in Suitland, Maryland since 1942. A new headquarters complex was completed in 2007 and supports over 4,000 employees.[15] The Bureau operates regional offices in 12 cities: Boston, New York City, Philadelphia, Detroit, Chicago, Kansas City, Seattle, Charlotte, Atlanta, Dallas, Denver, and Los Angeles. The National Processing Center is located in Jeffersonville, Indiana. Additional temporary processing facilities are used to facilitate the decennial census, which employs more than a million people. The cost of the 2000 Census was $4.5 billion. During the years just prior to the decennial census, parallel census offices, known as "Regional Census Centers" are opened in the field office cities. The decennial operations are carried out from these facilities. The Regional Census Centers l oversee the openings and closings of smaller "Local Census Offices" within their collection jurisdictions. The estimated cost of the 2010 Census is $14.7 billion.

The Census Bureau also runs the Census Information Center cooperative program that involves 58 "national, regional, and local non-profit organizations." The CIC program aims to represent the interests of underserved communities.[16]

Census regions and divisions

The United States Census Bureau has four official regions, with nine official divisions.

- Northeastern United States
  - New England
  - Middle Atlantic
- Midwestern United States
  - East North Central States
  - West North Central States
- Western United States
  - Pacific States
  - Mountain States
- Southern United States
  - West South Central States
  - East South Central States
  - South Atlantic States

Computer equipment

The 1890 census was the first to use the electric tabulating machines invented by Herman Hollerith.[17] [18] For 1890-1940 details, see Truesdell, Leon E. (1965). *The Development of Punch Card Tabulation in the Bureau of the Census, 1890-1940: With outlines of actual tabulation programs*. US GPO. In 1946, knowing of the Bureau's funding of Hollerith and, later, Powers, John Mauchly approached the Bureau about early funding for UNIVAC development.[19] A UNIVAC I computer was accepted by the Bureau in 1951.[20]

In order to reduce paper usage and reduce payroll expenses, 500,000 handheld computers (HHC's) were used for the first time in 2009 during the address canvassing portion of the 2010 Decennial Census Project. Projected savings are over one billion dollars.[21] The further implementation of advanced computer systems in the planning for the 2020 census provides major challenges for an upgrade in the technological broadening of census protocol.

Ongoing surveys

A survey is a method of collecting and analyzing social, economic, and geographic data. It provides information about the conditions of the United States, states, and counties. Throughout the decade between censuses, the bureau is conducts surveys to produce a general view and comprehensive study of the United States' social and economic conditions.

Staff from the Current Surveys Program conduct ongoing and special surveys about people and their characteristics. A network of professional field representatives gathers information from a sample of households, responding to questions about employment, consumer expenditures, health, housing, and other topics.

Surveys conducted between decades:

- American Community Survey
- Consumer Expenditure Survey
- Census of Governments
- Current Population Survey
- Economic Census
- National Hospital Discharge Survey
- National Health Interview Survey
- National Hospital Ambulatory Medical Care Survey
- National Crime Victimization Survey
- National Ambulatory Medical Care Survey (NAMCS)
- National Nursing Home Survey
- Survey of Income and Program Participation
- Survey of Construction
- Survey of Market Absorption
- Survey of Program Dynamics

- National Longitudinal Survey
- National Survey of Fishing, Hunting, & Wildlife-Associated Recreation 2001
- American Housing Survey
- Residential Finance Survey
- National Epidemiologic Survey of Alcohol Related Conditions

Other surveys conducted

The Census Bureau collects information in many other surveys and provides the data to the survey sponsor for release. These sponsors include:

- Bureau of Justice Statistics (BJS)
- Bureau of Labor Statistics (BLS)
- Bureau of Transportation Statistics (BTS)
- Department of Housing and Urban Development (HUD)
- National Center for Education Statistics (NCES)
- National Center for Health Statistics (NCHS)
- National Science Foundation (NSF)
- Social Security Administration (SSA)

Notes

[1] http://www.census.gov

[2] http://www.law.cornell.edu/uscode/13/1.html

[3] USCB DOC-D1026 QVC Manual 01/03/09

[4] "Census Population Estimates" (http://www.census.gov/popest/estimates.php). U.S. Bureau of the Census. .

[5] History 1790 (http://www.census.gov/history/www/overview/012270.html). US Census Bureau.

[6] History 1840 (http://www.census.gov/history/www/overview/012275.html). US Census Bureau.

[7] History: 1900 Overview (http://www.census.gov/history/www/through_the_decades/overview/1900.html). US Census Bureau.

[8] History 1920 (http://www.census.gov/history/www/overview/012283.html). US Census Bureau.

[9] History 1954 (http://www.census.gov/history/www/overview/012286.html). US Census Bureau.

[10] *Census Employee Handbook* (http://www.census.gov/po/www/foia/D590.pdf), April 2009,

[11] *The Myth of Census Confidentiality* (http://aascpress.metapress.com/content/jk115283m54v4313/), **8**, Amerasia Journal, UCLA Asian American Studies Center Press, Fall/Winter 1981, pp. 111–120, ISSN 0044-7471,

[12] David Kopel (May 4, 1990), *Census Confidentiality? The Check's is in the Mail* (http://www.cato.org/pub_display.php?pub_id=8938), Cato Institute,

[13] JR Minkel (March 30, 2007), *Confirmed: The U.S. Census Bureau Gave Up Names of Japanese-Americans in WW II* (http://www.scientificamerican.com/article.cfm?id=confirmed-the-us-census-b&sc=I100322), Scientific American,

[14] Haya El Nasser (March 30, 2007), *Papers show Census role in WWII camps* (http://www.usatoday.com/news/nation/2007-03-30-census-role_N.htm), USA Today,

[15] Census.gov (http://www.census.gov/aboutus/who.html)

[16] "Census Information Centers" (http://www.census.gov/clo/www/cic/). U.S. Bureau of the Census. . Retrieved 2008-05-13.

[17] Herman Hollerith (http://www.census.gov/history/www/notable_alumni/010944.html)

[18] History 1890 (http://www.census.gov/history/www/overview/012280.html) US Census Bureau.

[19] Stern, Nancy (1981). *From ENIAC to UNIVAC: An appraisal of the Eckert-Mauchly Computers*. Digital Press. ISBN 0-932376-14-2.

[20] Bashe, Charles J.; et. al (1986). *IBM's Early Computers*. MIT. ISBN 0-262-02225-7.

[21] Govcomm.harris.com (http://www.govcomm.harris.com/solutions/products/000138.asp)

References

- The original version of, and subsequent additions to, this article were adapted from U.S. Census Bureau (http://www.census.gov/acsd/www/history.html) text.
- Geographic Areas Reference Manual (http://www.census.gov/geo/www/garm.html) from the U.S. Census Bureau contains detailed explanations of geographic terms used in the census.
- The mission and operation of the U.S. Census Bureau are laid out in Title 13 of the United States Code.

External links

- Official website (http://http://www.census.gov)
- U.S. and World Population Clocks – POPClocks (http://www.census.gov/main/www/popclock.html)
- Works by Bureau of the Census (http://www.gutenberg.org/author/United_States._Bureau_of_the_Census) at Project Gutenberg
- Census Bureau Meeting Notices and Rule Changes (http://thefederalregister.com/b.p/agency/Census_Bureau/) from The Federal Register RSS Feed (http://thefederalregister.com/rss/agency/Census_Bureau/)
- Proposed and finalized federal regulations from the United States Census Bureau (http://openregs.com/agencies/view/173/bureau_of_the_census)

Population density

Population density (in agriculture **standing stock** and standing crop) is a measurement of population per unit area or unit volume. It is frequently applied to living organisms, and particularly to humans. It is a key geographic term.[1]

Population density (people per km^2) by country, 2006

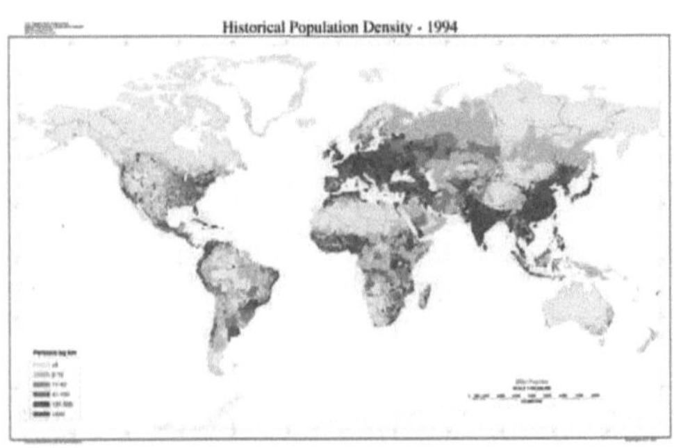

Population density (people per km^2) map of the world in 1994 (detailed).

Biological population densities

Population density is population divided by total land area.[1]

Low densities may cause an extinction vortex and lead to further reduced fertility. This is called the Allee effect after the scientist who identified it. Examples of the causes in low population densities include:[2]

- Increased problems with locating mates
- Increased inbreeding

Different species have different expected densities. R-selected species commonly have high population densities, while

K-selected species may have lower densities.[3] Low densities may be associated with specialized mate location adaptations such as specialized pollinators, as found in the orchid family (*Orchidaceae*).

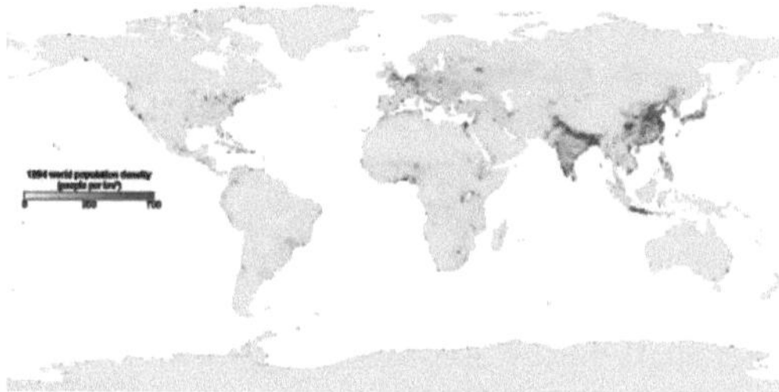

Population density (people per km^2) map of the world in 1994.

Human population density

For humans, population density is the number of people per unit of area usually per square kilometer or mile (which may include or exclude cultivated or potentially productive area). Commonly this may be calculated for a county, city, country, another territory, or the entire world.

Sai Yeung Choi Street South in Mong Kok, Hong Kong, one of the most densely populated places in the world

The world's population is 6.8 billion,[4] and Earth's total area (including land and water) is 510 million square kilometers (197 million square miles).[5] Therefore the worldwide human population density is 6.8 billion ÷ 510 million = 13.3 per km² (34.5 per sq. mile). If only the Earth's land area of 150 million km² (58 million sq. miles) is taken into account, then human population density increases to 45.3 per km² (117.2 per sq. mile). This calculation includes all continental and island land area, including Antarctica. If Antarctica is also excluded, then population density rises to 50 people per km² (129.28 per sq. mile).[1] Considering that over half of the Earth's land mass consists of areas inhospitable to human inhabitation, such as deserts and high mountains, and that population tends to cluster around seaports and fresh water sources, this number by itself does not give any meaningful measurement of human population density.

Mongolia is the least densely populated country in the world; Monaco is currently the most.

Several of the most densely-populated territories in the world are city-states, microstates, micronations, or dependencies.[6] [7] These territories share a relatively small area and a high urbanization level, with an economically specialized city population drawing also on rural resources outside the area, illustrating the difference between high population density and overpopulation.

Cities with high population densities are, by some, considered to be overpopulated, though the extent to which this is the case depends on factors like quality of housing and infrastructure and access to resources.[8] Most of the most densely-populated cities are in southern and eastern Asia, though Cairo and Lagos in Africa also fall into this category.[9]

City population is, however, heavily dependent on the definition of "urban area" used: densities are often higher for the central municipality itself, than when more recently-developed and administratively separate suburban communities are included, as in the concepts of agglomeration or metropolitan area, the latter including sometimes neighboring cities. For instance, Milwaukee has a greater population density when just the inner city is measured, and not the surrounding suburbs as well.[10]

Other methods of measurement

While arithmetic density is the most common way of measuring population density, several other methods have been developed which aim to provide a more accurate measure of population density over a specific area.

- **Arithmetic density**: The total number of people / area of land (measured in km^2 or sq miles).
- **Physiological density**: The total population / area of arable land.
- **Agricultural density**: The total *rural* population / area of arable land.
- **Residential density** : The number of people living in an urban area / area of residential land.
- **Urban density** : The number of people inhabiting an urban area / total area of urban land.
- **Ecological optimum**: The density of population which can be supported by the natural resources.

See also

- Human geography
- Idealized population
- Optimum population
- Population bottleneck
- Population genetics
- Population health
- Population momentum
- Population pyramid
- Rural transport problem
- Small population size

Lists:

- List of cities by population
- List of cities by population density
- List of European cities proper by population density
- List of islands by population density
- List of countries by population density
- List of U.S. states by population density

References

[1] About.com (http://geography.about.com/od/populationgeography/a/popdensity.htm)
[2] Minimum viable population size (http://www.eoearth.org/article/Minimum_viable_population_size)
[3] Density-Dependent Selection (http://science.jrank.org/pages/48388/Density-Dependent-Selection.html)
[4] (http://www.census.gov/main/www/popclock.html)
[5] (https://www.cia.gov/library/publications/the-world-factbook/geos/xx.html)
[6] Department of Economic and Social Affairs Population Division (2009) (.PDF). *World Population Prospects, Table A.1* (http://www.un.org/esa/population/publications/wpp2008/wpp2008_text_tables.pdf). 2008 revision. United Nations. . Retrieved 2009-03-12.
[7] The Monaco government uses a smaller surface area figure resulting in a population density of 18,078 per km^2
[8] Human Population - Global Issues (http://www.globalissues.org/issue/198/human-population)
[9] The largest cities in the world by land area, population and density (http://www.citymayors.com/statistics/largest-cities-density-125.html)
[10] The Population of Milwaukee County (http://www.wisconline.com/greenmap/milwaukee/population.html)

External links

- Selected Current and Historic City, Ward & Neighborhood Densities (http://www.demographia.com/db-citydenshist.htm)

pfl:B'velgerungsdischd

Marriage

Marriage (or **wedlock**) is a social union or legal contract between people that creates kinship. It is an institution in which interpersonal relationships, usually intimate and sexual, are acknowledged in a variety of ways, depending on the culture or subculture in which it is found. Such a union, often formalized via a wedding ceremony, may also be called *matrimony*.

People marry for many reasons, including one or more of the following: legal, social, libidinal, emotional, economic, spiritual, and religious. These might include arranged marriages, family obligations, the legal establishment of a nuclear family unit, the legal protection of children and public declaration of commitment.[1] [2] The act of marriage usually creates normative or legal obligations between the individuals involved. In some societies these obligations also extend to certain family members of the married persons. Some cultures allow the dissolution of marriage through divorce or annulment.

Marriage is usually recognized by the state, a religious authority, or both. It is often viewed as a contract. Civil marriage is the legal concept of marriage as a governmental institution irrespective of religious affiliation, in accordance with marriage laws of the jurisdiction.

Definitions

> “Marriage is the union of two different surnames, in friendship and in love, in order to continue the posterity of the former sages, and to furnish those who shall preside at the sacrifices to heaven and earth, at those in the ancestral temple, and at those at the altars to the spirits of the land and grain.”
>
> —Confucius, [3]

Anthropologists have proposed several competing definitions of marriage so as to encompass the wide variety of marital practices observed across cultures.[4] In his book *The History of Human Marriage* (1921), Edvard Westermarck defined marriage as "a more or less durable connection between male and female lasting beyond the mere act of propagation till after the birth of the offspring."[5] In *The Future of Marriage in Western Civilization* (1936), he rejected his earlier definition, instead provisionally defining marriage as "a relation of one or more men to one or more women that is recognised by custom or law".[6]

The anthropological handbook *Notes and Queries* (1951) defined marriage as "a union between a man and a woman such that children born to the woman are the recognized legitimate offspring of both partners."[7] In recognition of a practice by the Nuer of Sudan allowing women to act as a husband in certain circumstances, Kathleen Gough suggested modifying this to "a woman and one or more other persons."[8]

Edmund Leach criticized Gough's definition for being too restrictive in terms of recognized legitimate offspring and suggested that marriage be viewed in terms of the different types of rights it serves to establish. Leach expanded the definition and proposed that "Marriage is a relationship established between a woman and one or more other persons, which provides that a child born to the woman under circumstances not prohibited by the rules of the relationship, is accorded full birth-status rights common to normal members of his society or social stratum"[9] Leach argued that no one definition of marriage applied to all cultures. He offered a list of ten rights associated with marriage, including sexual monopoly and rights with respect to children, with specific rights differing across cultures.[10]

Duran Bell also criticized the legitimacy-based definition on the basis that some societies do not require marriage for legitimacy, arguing that in societies where illegitimacy means only that the mother is unmarried and has no other legal implications, a legitimacy-based definition of marriage is circular. He proposed defining marriage in terms of sexual access rights.[4]

Etymology

The modern English word "marriage" derives from Middle English *mariage*, which first appears in 1250–1300 C.E. This in turn is derived from Old French *marier* (to marry) and ultimately Latin *marītāre* (to marry) and *marītus* (of marriage).[11]

History of marriage by culture

A pair of wedding rings

Although the institution of marriage pre-dates reliable recorded history, many cultures have legends concerning the origins of marriage. The way in which a marriage is conducted and its rules and ramifications has changed over time, as has the institution itself, depending on the culture or demographic of the time.[12] Various cultures have had their own theories on the origin of marriage. One example may lie in a man's need for assurance as to paternity of his children. He might therefore be willing to pay a bride price or provide for a woman in exchange for exclusive sexual access.[13] Legitimacy is the consequence of this transaction rather than its motivation. In Comanche society, married women work harder, lose sexual freedom, and do not seem to obtain any benefit from marriage.[14] But nubile women are a source of jealousy and strife in the tribe, so they are given little choice other than to get married. "In almost all societies, access to women is institutionalized in some way so as to moderate the intensity of this competition."[15] Forms of group marriage which involve more than one member of each sex, and therefore are not either polygyny or polyandry, have existed in history. However, these forms of marriage are extremely rare. Of the 250 societies reported by the American anthropologist George P. Murdock in 1949, only the Caingang of Brazil had any group marriages at all.[16]

Various marriage practices have existed throughout the world. In some societies an individual is limited to being in one such couple at a time (*monogamy*), while other cultures allow a male to have more than one wife (*polygyny*) or, less commonly, a female to have more than one husband (*polyandry*). Some societies also allow marriage between two males or two females. Societies frequently have other restrictions on marriage based on the ages of the participants, pre-existing kinship, and membership in religious or other social groups.

Europe

In Ancient Greece, no specific civil ceremony was required for the creation of a marriage – only mutual agreement and the fact that the couple must regard each other as husband and wife accordingly. Men usually married when they were in their 20s or 30s and expected their wives to be in their early teens. It has been suggested that these ages made sense for the Greeks because men were generally done with military service by age 30, and marrying a young girl ensured her virginity. Married Greek women had few rights in ancient Greek society and were expected to take care of the house and children. Time was an important factor in Greek marriage. For example, there were superstitions that being married during a full moon was good luck and, according to Robert Flacelière, Greeks married in the winter. Inheritance was more important than feelings: A woman whose father dies without male heirs can be forced to marry her nearest male relative—even if she has to divorce her husband first.[17]

There were several types of marriages in ancient Roman society. The traditional ("conventional") form called *conventio in manum* required a ceremony with witnesses and was also dissolved with a ceremony.[18] In this type of marriage, a woman lost her family rights of inheritance of her old family and gained them with her new one. She now was subject to the authority of her husband. There was the free marriage known as *sine manu*. In this arrangement, the wife remained a member of her original family; she stayed under the authority of her father, kept her family rights of inheritance with her old family and did not gain any with the new family.[19] The minimum age of marriage for girls was 12.[20]

From the early Christian era (30 to 325 CE), marriage was thought of as primarily a private matter, with no uniform religious or other ceremony being required. However, bishop Ignatius of Antioch writing around 110 to bishop Polycarp of Smyrna exhorts, "[I]t becomes both men and women who marry, to form their union with the approval of the bishop, that their marriage may be according to God, and not after their own lust."[21]

Woodcut. How Reymont and Melusina were betrothed / And by the bishop were blessed in their bed on their wedlock. From the *Melusine*, 15th century.

In the 12th century women were obligated to take the name of their husbands and starting in the second half of the 16th century parental consent along with the church's consent was required for marriage.[22]

With few local exceptions, until 1545, Christian marriages in Europe were by mutual consent, declaration of intention to marry and upon the subsequent physical union of the parties.[23] [24] The couple would promise verbally to each other that they would be married to each other; the presence of a priest or witnesses was not required.[25] This promise was known as the "verbum." If freely given and made in the present tense (e.g., "I marry you"), it was unquestionably binding;[23] if made in the future tense ("I will marry you"), it would constitute a betrothal. One of the functions of churches from the Middle Ages was to register marriages, which was not obligatory. There was no state involvement in marriage and personal status, with these issues being adjudicated in ecclesiastical courts. During the Middle Ages marriages were arranged, sometimes as early as birth, and these early pledges to marry were often used to ensure treaties between different royal families, nobles, and heirs of fiefdoms. The church resisted these imposed unions, and increased the number of causes for nullification of these arrangements.[22] As Christianity spread during the Roman period and the Middle Ages, the idea of free choice in selecting marriage partners increased and spread with it.[22]

The average age of marriage in the late 13th century into the 16th century was around 25 years of age.[26]

As part of the Protestant Reformation, the role of recording marriages and setting the rules for marriage passed to the state, reflecting Martin Luther's view that marriage was a "worldly thing".[27] By the 17th century many of the Protestant European countries had a state involvement in marriage. As of 2000, the average marriage age range was

25–44 years for men and 22–39 years for women. In England, under the Anglican Church, marriage by consent and cohabitation was valid until the passage of Lord Hardwicke's Act in 1753. This act instituted certain requirements for marriage, including the performance of a religious ceremony observed by witnesses.[28]

As part of the Counter-Reformation, in 1563 the Council of Trent decreed that a Roman Catholic marriage would be recognized only if the marriage ceremony was officiated by a priest with two witnesses. The Council also authorized a Catechism, issued in 1566, which defined marriage as, "The conjugal union of man and woman, contracted between two qualified persons, which obliges them to live together throughout life."[29]

In the early modern period, John Calvin and his Protestant colleagues reformulated Christian marriage by enacting the Marriage Ordinance of Geneva, which imposed "The dual requirements of state registration and church consecration to constitute marriage"[29] for recognition.

In England and Wales, Lord Hardwicke's Marriage Act 1753 required a formal ceremony of marriage, thereby curtailing the practice of Fleet Marriage.[30] These were clandestine or irregular marriages performed at Fleet Prison, and at hundreds of other places. From the 1690s until the Marriage Act of 1753 as many as 300,000 clandestine marriages were performed at Fleet Prison alone.[31] The Act required a marriage ceremony to be officiated by an Anglican priest in the Anglican Church with two witnesses and registration. The Act did not apply to Jewish marriages or those of Quakers, whose marriages continued to be governed by their own customs.

In England and Wales, since 1837, civil marriages have been recognized as a legal alternative to church marriages under the Marriage Act of 1836. In Germany, civil marriages were recognized in 1875. This law permitted a declaration of the marriage before an official clerk of the civil administration, when both spouses affirm their will to marry, to constitute a legally recognized valid and effective marriage, and allowed an optional private clerical marriage ceremony.

In contemporary English common law, a marriage is a voluntary contract by a man and a woman, in which by agreement they choose to become husband and wife.[32] Edvard Westermarck proposed that "the institution of marriage has probably developed out of a primeval habit".[33]

China

The mythological origin of Chinese marriage is a story about Nüwa and Fu Xi who invented proper marriage procedures after becoming married. In ancient Chinese society, people of the same surname were not supposed to marry and doing so was seen as incest. However, because marriage to one's maternal relatives was not thought of as incest, families sometimes intermarried from one generation to another. Over time, Chinese people became more geographically mobile. Individuals remained members of their biological families. When a couple died, the husband and the wife were buried separately in the respective clans' graveyard. In a maternal marriage, a male would become a son-in-law who lived in the wife's home.

The New Marriage Law of 1950 radically changed Chinese marriage traditions, enforcing monogamy, equality of men and women, and choice in marriage; arranged marriages were the most common type of marriage in China until then.

Same-sex marriage

Various types of same-sex marriages have existed,[34] ranging from informal, unsanctioned relationships to highly ritualized unions.[35]

While it is a relatively new practice that same-sex couples are being granted the same form of legal marital recognition as commonly used by mixed-sex couples, recent publicity and debate over the past decade gives an impression that civil marriage for lesbian and gay couples is novel and untested. There is a long history of recorded same-sex unions around the world.[36] It is believed that same-sex unions were celebrated in Ancient Greece and Rome,[36] some regions of China, such as Fujian, and at certain times in ancient European history.[37] A law in the

Theodosian Code (*C. Th.* 9.7.3) issued in 342 CE imposed severe penalties or death on same-sex marriage in ancient Rome[38] but the exact intent of the law and its relation to social practice is unclear, as only a few examples of same-sex marriage in that culture exist.[39]

Group marriage

Group marriage also known as multi-lateral marriage, is a form of polyamory in which more than two persons form a family unit, with all the members of the group marriage being considered to be married to all the other members of the group marriage, and all members of the marriage share parental responsibility for any children arising from the marriage.[40] No country legally condones group marriages, neither under the law nor as a common law marriage.

Selection of a partner

An arranged marriage between Louis XIV of France and Maria Theresa of Spain.

There is wide cross-cultural variation in the social rules governing the selection of a partner for marriage. There is variation in the degree to which partner selection is an individual decision by the partners or a collective decision by the partners' kin groups, and there is variation in the rules regulating which partners are valid choices.

In many societies the choice of partner is limited to suitable persons from specific social groups. In some societies the rule is that a partner is selected from an individual's own social group – endogamy, this is the case in many class and caste based societies. But in other societies a partner must be chosen from a different group than one's own – exogamy, this is the case in many societies practicing totemic religion where society is divided into several exogamous totemic clans, such as most Aboriginal Australian societies. In other societies a person is expected to marry their cross-cousin, a woman must marry her father's sister's son and a man must marry his mother's brother's daughter – this is often the case if either a society has a rule of tracing kinship exclusively through patrilineal or matrilineal descent groups as among the Akan people of Africa. Another kind of marriage selection is the levirate marriage in which widows are obligated to marry their husband's brother, this is mostly found in societies where kinship is based on endogamous clan groups.

In other cultures with less strict rules governing the groups from which a partner can be chosen the selection of a marriage partner may involve either the couple going through a selection process of courtship or the marriage may be arranged by the couple's parents or an outside party, a matchmaker.

A pragmatic (or 'arranged') marriage is made easier by formal procedures of family or group politics. A responsible authority sets up or encourages the marriage; they may, indeed, engage a professional matchmaker to find a suitable spouse for an unmarried person. The authority figure could be parents, family, a religious official, or a group consensus. In some cases, the authority figure may choose a match for purposes other than marital harmony.

In rural Indian villages, child marriage is also practiced, with parents at times arranging the wedding, sometimes even before the child is born.[41] This practice was made illegal under the Child Marriage Restraint Act of 1929.

In some societies ranging from Central Asia to the Caucasus to Africa, the custom of bride kidnapping still exists, in which a woman is captured by a man and his friends. Sometimes this covers an elopement, but sometimes it depends on sexual violence. In previous times, *raptio* was a larger-scale version of this, with groups of women captured by groups of men, sometimes in war; the most famous example is The Rape of the Sabine Women, which provided the first citizens of Rome with their wives.

Other marriage partners are more or less imposed on an individual. For example, widow inheritance provides a widow with another man from her late husband's brothers.

Marriage ceremony

A marriage is usually formalized at a wedding or marriage ceremony. The ceremony may be officiated either by a religious official, by a government official or by a state approved celebrant. In many European and some Latin American countries, any religious ceremony must be held separately from the required civil ceremony. Some countries – such as Belgium, Bulgaria, France, the Netherlands, Romania and Turkey[42] – require that a civil ceremony take place before any religious one. In some countries – notably the United States, Canada, the United Kingdom, the Republic of Ireland, Norway and Spain – both ceremonies can be held together; the officiant at the religious and civil ceremony also serving as agent of the state to perform the civil ceremony. To avoid any implication that the state is "recognizing" a religious marriage (which is prohibited in some countries) – the "civil" ceremony is said to be taking place at the same time as the religious ceremony. Often this involves simply signing a register during the religious ceremony. If the civil element of the religious ceremony is omitted, the marriage is not recognized by government under the law.

Couple married in a Shinto ceremony in Takayama, Gifu prefecture.

While some countries, such as Australia, permit marriages to be held in private and at any location, others, including England and Wales, require that the civil ceremony be conducted in a place open to the public and specially sanctioned by law. In England, the place of marriage need no longer be a church or register office, but could also be a hotel, historic building or other venue that has obtained the necessary license. An exception can be made in the case of marriage by special emergency license, which is normally granted only when one of the parties is terminally ill. Rules about where and when persons can marry vary from place to place. Some regulations require that one of the parties reside in the locality of the registry office.

Within the parameters set by the law of the jurisdiction in which a marriage or wedding takes place, each religious authority has rules for the manner in which weddings are to be conducted by their officials and members.

Cohabitation

Marriage is an institution which can join together people's lives in a variety of emotional and economic ways. In many Western cultures, marriage usually leads to the formation of a new household comprising the married couple, with the married couple living together in the same home, often sharing the same bed, but in some other cultures this is not the tradition.[43] Among the Minangkabau of West Sumatra, residency after marriage is matrilocal, with the husband moving into the household of his wife's mother.[44] Residency after marriage can also be patrilocal or avunculocal. Such marriages have also been increasingly common in modern Beijing. Guo Jianmei, director of the center for women's studies at Beijing University, told a Newsday correspondent, "Walking marriages reflect sweeping changes in Chinese society."[45] A similar arrangement in Saudi Arabia, called misyar marriage, also involves the husband and wife living separately but meeting regularly.[46]

Conversely, marriage is not a prerequisite for cohabitation. In some cases couples living together do not wish to be recognized as married, such as when pension or alimony rights are adversely affected, or because of taxation consideration, or because of immigration issues, and for many other reasons. In modern western societies some couples cohabitate before marriage to test whether such an arrangement might work in the long term.

In some cases cohabitation may constitute a common-law marriage, and in some countries the laws recognize cohabitation in preference to the formality of marriage for taxation and social security benefits. This is the case, for example, in Australia.[47]

Sex and procreation

Many of the world's major religions look with disfavor on sexual relations outside of marriage.[48] Many non-secular states, mostly with Muslim majorities, sanction criminal penalties for sexual intercourse before marriage. Sexual relations by a married person with someone other than his/her spouse is known as adultery and is also frequently disapproved by the major world religions (some calling it a sin). Adultery is considered in many jurisdictions to be a crime and grounds for divorce.

Christ and the woman taken in adultery by Jan Brueghel the Elder, Pinakothek

On the other hand, marriage is not a prerequisite for having children. In the United States, the National Center for Health Statistics reported that in 1992, 30.1 percent of births were to unmarried women.[49] [50] In 2006, that number had risen to 38.5 percent.[51] Children born outside of marriage, bastards and whoresons, were known as illegitimate and suffered legal disadvantages and social stigma. In recent years the legal relevance of illegitimacy has declined and social acceptance has increased, especially in western countries.

Some married couples choose not to have children and so remain childfree. Others are unable to have children because of infertility or other factors preventing conception or the bearing of children. In some cultures, marriage imposes an *obligation* on women to bear children. In northern Ghana, for example, payment of bridewealth signifies a woman's requirement to bear children, and women using birth control face substantial threats of physical abuse and reprisals.[52]

Marriage law

Marriage laws refer to the legal requirements which determine the validity of a marriage, which vary considerably between countries.

Common-law marriage

In some jurisdictions but not all, marriage relationships may be created by the operation of the law alone. Unlike the typical ceremonial marriage with legal contract, wedding ceremony, and other details, a common-law marriage may be called "marriage by habit and repute (cohabitation)." A **de facto** common-law marriage without a license or ceremony is legally binding in some jurisdictions but has no legal consequence in others.[53]

Rights and obligations

A marriage bestows rights and obligations on the married parties, and sometimes on relatives as well, being the sole mechanism for the creation of affinal ties (in-laws). These may include:

- Giving a husband/wife or his/her family control over a spouse's sexual services, labor, and property.
- Giving a husband/wife responsibility for a spouse's debts.
- Giving a husband/wife visitation rights when his/her spouse is incarcerated or hospitalized.
- Giving a husband/wife control over his/her spouse's affairs when the spouse is incapacitated.
- Establishing the second legal guardian of a parent's child.
- Establishing a joint fund of property for the benefit of children.
- Establishing a relationship between the families of the spouses.

A Ketubah in Hebrew, a Jewish marriage-contract outlining the duties of each partner.

These rights and obligations vary considerably between societies, and between groups within society.[54]

Marriage restrictions

Marriage is an institution that is historically filled with restrictions. From age, to race, to social status, to consanguinity, to gender, restrictions are placed on marriage by society for reasons of benefiting the children, passing on healthy genes, maintaining cultural values, or because of prejudice and fear. Almost all cultures that recognize marriage also recognize adultery as a violation of the terms of marriage.[55]

The United States has had a history of marriage restriction laws. Many states enacted miscegenation laws which were first introduced in the late 17th century in the slave-holding colonies of Virginia (1691) and Maryland (1692) and lasted until 1967 (until it was overturned via *Loving v. Virginia*). Many of these states restricted several minorities from marrying whites. For example, Alabama, Arkansas, and Oklahoma banned Blacks in particular. States such as Mississippi and Missouri banned Blacks and Asians. States such as North Carolina and South Carolina banned Blacks and Native Americans, and some states such as Georgia, South Carolina, and Virginia banned all non-whites.

It is a relatively new practice that same-sex couples are being granted the same form of legal marital recognition available to mixed-sexed couples. In the United States, the 1996 Defense of Marriage Act (DOMA) explicitly defines marriage for the purposes of federal law as between a man and a woman and allows states to ignore same-sex marriages from other states (though states arguably could do this already).[56] [57] Forty-one US states currently define marriage as between a man and a woman. Three of those states have statutory language that pre-dates DOMA (enacted before 1996) defining marriage as such. Thirty states have defined marriage in their constitutions. Arizona is the only state that has ever defeated a constitutional amendment defining marriage as only between a man and a woman (2006), but it subsequently passed one in 2008.[58]

Societies have often placed restrictions on marriage to relatives, though the degree of prohibited relationship varies widely. With few exceptions, marriages between parents and children or between full siblings have been considered incest and forbidden. However, marriages between more distant relatives have been much more common, with one estimate being that 80% of all marriages in history have been between second cousins or closer.[59] In modern times this proportion has fallen dramatically, but still more than 10% of all marriages are believed to be between first and second cousins.[60] In the United States, such marriages are now highly stigmatized, and laws ban most or all first-cousin marriage in 30 states. Specifics vary: in South Korea, historically it was illegal to marry someone with the same last name.[61]

Many societies have required a person to marry within their own general social group, which anthropologists refer to as endogamy. An example of such restrictions would be a requirement to marry someone from the same tribe.

Restrictions against polygamy have been common. Opposition to the recognition of Deseret as a State by the Federal government was founded on opposition to the once-practiced polygamous marriages of Mormons.

State recognition

In many jurisdictions, a civil marriage may take place as part of the religious marriage ceremony, although they are theoretically distinct. Some jurisdictions allow civil marriages in circumstances which are notably not allowed by particular religions, such as same-sex marriages or civil unions.

Marriage and religion

Among the precepts of mainstream religions are found, as a rule, unequivocal prescriptions for marriage, establishing both rituals and rules of conduct.

Abrahamic religions

In the Hebrew Bible (Old Testament)

The Hebrew Bible (Christian Old Testament) describes a number of marriages, including those of Isaac,[Gen 24:49–67] Jacob,[Gen 29:27] and Samson.[Judg 14:7–12] Polygyny, or men having multiple wives at once, is one of the most common marital arrangements represented in the Hebrew Bible.[62]

Rembrandt's depiction of Samson's marriage feast

Betrothal (*erusin*), which is merely a binding promise to get married, is distinct from marriage itself (*nissu'in*), with the time between these events varying substantially.[62] [63] Since a wife was regarded as property in those days, the betrothal (*erusin*) was effected simply by purchasing her from her father (or guardian);[62] [63] the girl's consent is not explicitly required by any biblical law.[63] Like the adjacent Arabic culture (in the pre-Islamic period),[64] the act of marriage appears mainly to have consisted of the groom fetching the bride, although among the Israelites (unlike the Arabs) the procession was a festive occasion, accompanied by music, dancing, and lights.[62] [63] To celebrate the marriage, week-long feasts were sometimes held.[62] [63]

A Jewish wedding, painting by Jozef Israëls, 1903.

In biblical times, a wife was regarded as chattel, belonging to her husband;[62] [63] the descriptions of the bible suggest that she would be expected to perform tasks such as spinning, sewing, weaving, manufacture of clothing, fetching of water, baking of bread, and animal husbandry.[65] However, wives were usually looked after with care, and men with more than one wife were expected to ensure that they continue to give the first wife food, clothing, and marital rights.[Ex 21:10]

Since a wife was regarded as property, her husband was originally free to divorce her for any reason, at any time.[63] A divorced couple were permitted to get back together, unless the wife had married someone else after her divorce.[Deut 24:2–4]

Judaism

In Judaism, marriage is viewed as a contractual bond commanded by God in which a man and a woman come together to create a relationship in which God is directly involved.[Deut. 24:1] Though procreation is not the sole purpose, a Jewish marriage is also expected to fulfill the commandment to have children.[Gen. 1:28] The main focus centers around the relationship between the husband and wife. Kabbalistically, marriage is understood to mean that the husband and wife are merging together into a single soul. This is why a man is considered "incomplete" if he is not married, as his soul is only one part of a larger whole that remains to be unified.[66]

Christianity

Christians variously regard marriage as a sacrament, a contract, a sacred institution, or a covenant.[67] From the very beginning of the Christian Church, marriage law and theology have been a major matter.[68] The foundation of the Western tradition of Christian marriages have been the teachings of Jesus Christ and the Apostle Paul.[29]

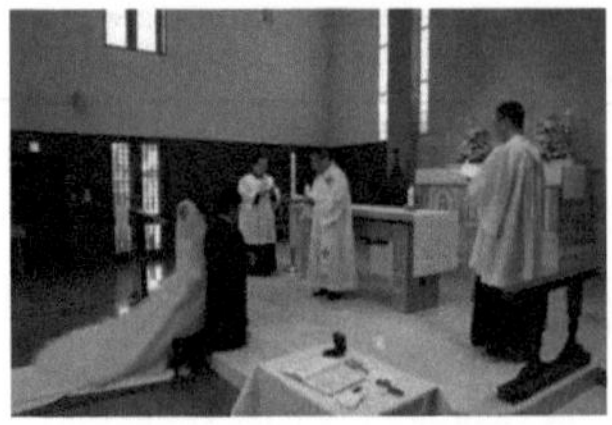

Christian wedding in Kyoto, Japan.

Christians often marry for religious reasons ranging from following the biblical injunction for a "man to leave his father and mother and cleave to his wife, and the two shall become one,"[Gen. 2:24][69] to obeying Canon Law stating marriage between baptized persons is a sacrament.[70]

Divorce and remarriage while generally not encouraged are regarded differently by each Christian denomination. Most Protestant Churches allow people to marry again after a divorce. The Eastern Orthodox Church allows divorce for a limited number of reasons, and in theory, but usually not in practice, requires that a marriage after divorce be celebrated with a penitential overtone. In the Roman Catholic Church, marriage can only be ended by an annulment where the Church for special reasons regards it as never having taken place.[71]

> "'...So they are no longer two, but one. Therefore what God has joined together, let man not separate."
>
> — Jesus[Matthew 19:6]

Liturgical Christianity

Further information: Marriage in the Eastern Orthodox Church

Anglicans, Catholics, and Eastern Orthodox consider marriage termed *holy matrimony* to be an expression of divine grace, termed a *sacrament* or *mystery*. In Western ritual, the ministers of the sacrament are the husband and wife themselves, with a bishop, priest, or deacon merely witnessing the union on behalf of the church, and adding a blessing. In Eastern ritual churches, the bishop or priest functions as the actual minister of the Sacred Mystery (Eastern Orthodox deacons may not perform marriages). Western Christians commonly refer to marriage as a vocation, while Eastern Christians consider it an ordination and a martyrdom, though the theological emphases indicated by the various names are not excluded by the teachings of either tradition. Marriage is commonly celebrated in the context of a Eucharistic service (a nuptial Mass or Divine Liturgy). The sacrament of marriage is indicative of the relationship between Christ and the Church.[Eph. 5:29–32]

Roman Catholicism

The Roman Catholic tradition of the twelfth and thirteenth centuries defined marriage as a sacrament ordained by God,[29] signifying the mystical marriage of Christ to his Church.[72]

> "The matrimonial covenant, by which a man and a woman establish between themselves a partnership of the whole of life, is by its nature ordered toward the good of the spouses and the procreation and education of offspring; this covenant between baptized persons has been raised by Christ the Lord to the dignity of a sacrament."[73]

The mutual love between man and wife becomes an image of the eternal love with which God loves humankind. The celebration of marriage between two Catholics normally takes place during the public liturgical celebration of the Holy Mass, because of its sacramental connection with the unity of the Paschal mystery of Christ (Communion). Sacramental marriage confers a perpetual and exclusive bond between the spouses. By its nature, the institution of marriage and conjugal love is ordered to the procreation and upbringing of offspring. Marriage creates rights and duties in the Church between the spouses and towards their children: "[e]ntering marriage with the intention of never having children is a grave wrong and more than likely grounds for an annulment."[74]

According to current Catholic legislation governing marriage, "The essential properties of marriage are unity and indissolubility; in Christian marriage they acquire a distinctive firmness by reason of the sacrament.[75] Divorce is not recognized, but annulments predicated upon previously existing impediments may be granted. Offspring resulting from annulled relationships are considered legitimate. The remarriage of persons divorced from a living, lawful spouse are not separated from the Church, but they cannot receive Eucharistic communion.[76]

Protestantism

For Protestant denominations, the purposes of marriage include intimate companionship, rearing children and mutual support for both husband and wife to fulfill their life callings. Protestants generally approve of birth control and consider marital sexual pleasure to be a gift of God.

Most Reformed Christians would deny the elevation of marriage to the status of a sacrament, nevertheless it is considered a covenant between spouses before God.cf.[Ephesians 5:31–33]

Historically, five competing models of marriage in Christianity have shaped Western marriage and legal tradition:

- The Protestant Reformationists replaced the Roman Catholic sacramental model.
- Martin Luther saw it as a social "estate of the earthly kingdom...subject to the prince, not the Pope."
- John Calvin taught that marriage was a *covenant of grace* that required the coercive power of the state to preserve its integrity.
- Anglicans regarded marriage as a *domestic commonwealth* within England and the church. By the 17th century, Anglican theologians had begun to develop a theology of marriage to replace the sacramental model of marriage. These "regarded the interlocking commonwealths of state, church, and family as something of an earthly form of heavenly government."
- The secularism of the Enlightenment emphasized marriage as a *contract* "to be formed, maintained, and dissolved as the couple sees fit."[29]

Latter-day Saints

A couple following their marriage in the Manti Utah Temple.

Members of the Church of Jesus Christ of Latter-day Saints (LDS) believe that "marriage between a man and a woman is ordained of God and that the family is central to the Creator's plan for the eternal destiny of His children." The LDS belief is that marriage between a man and a woman can last beyond death and into eternity.[77]

Islam

Islam also commends marriage, with the age of marriage being whenever the individuals feel ready, financially and emotionally.

A Muslim bride of Pakistan origin signing the *nikkah nama* or marriage certificate.

In Islam, polygyny is allowed while polyandry is not, with the specific limitation that men can only have no more than four wives at any one time, with the requirement that they are able and willing to partition their time and wealth equally among the respective wives.

For a Muslim wedding to take place, the bride and her guardian must both agree on the marriage. Should either the guardian or the girl disagree on the marriage, it may not legally take place. In essence, while the guardian/father of the girl has no right to force her to marry, he has the right to stop a marriage from taking place, given that his reasons are valid. The professed purpose of this practice is to ensure that a woman finds a suitable partner whom she has chosen not out of sheer emotion.

From an Islamic (Sharia) law perspective, the minimum requirements and responsibilities in a Muslim marriage are that the groom provide living expenses (housing, clothing, food, maintenance) to the bride, and in return, the bride's main responsibility is raising children to be proper Muslims. All other rights and responsibilities are to be decided between the husband and wife, and may even be included as stipulations in the marriage contract before the marriage actually takes place, so long as they do not go against the minimum requirements of the marriage.

In Sunni Islam marriage must take place in the presence of at least two reliable witnesses, with the consent of the guardian of the bride and the consent of both the bride and the groom. Following the marriage, the couple may consummate the marriage. To create a religious contract between them, it is sufficient that a man and a woman indicate an intention to marry each other and recite the requisite words in front of a suitable Muslim, nowadays priest will be asked to officiate. The wedding party usually follows but can be held days, or months later, whenever the couple and their families want to, however there can be no concealment of the marriage as it is regarded as public notification due to the requirement of witnesses.[78] [79] [80] [81]

A Muslim couple being wed alongside the Tungabhadra River at Hampi, India.

In Shia Islam, marriage may take place without the presence of witnesses as is often the case in temporary mutta marriage (prohibited in Sunni Islam), but with the consent of both the bride and the groom. Following the marriage they may consummate their marriage.

Bahá'í

In the Bahá'í Faith marriage is encouraged and viewed as a mutually strengthening bond, but is not obligatory. A Bahá'í marriage requires the couple to choose each other, and then the consent of all living parents.[82]

Hinduism

Hinduism sees marriage as a sacred duty that entails both religious and social obligations. Old Hindu literature in Sanskrit gives many different types of marriages and their categorization ranging from "Gandharva Vivaha" (instant marriage by mutual consent of participants only, without any need for even a single third person as witness) to normal (present day) marriages, to "Rakshasa Vivaha" ("demoniac" marriage, performed by abduction of one participant by the other participant, usually, but not always, with the help of other persons). The Hindu Widow's Remarriage Act 1856 empowers a Hindu widow to remarry. Though traditionally widow remarriages were frowned upon and are still considered taboo in many parts of India,[83] the society is changing and the incidence of widow remarriage is on a rise.[84]

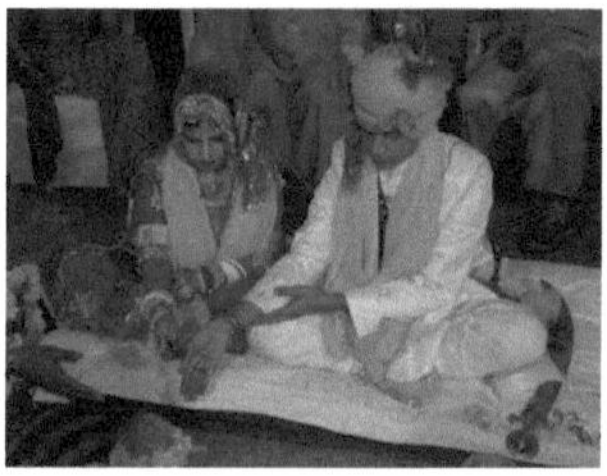
Hindu marriage ceremony from a Rajput wedding.

Although sati, or the practice of a widow immolating herself on her husband's funeral pyre, was officially outlawed by India's British rulers in 1829, the rite persists. The most high-profile sati incident was in Rajasthan in 1987 when 18-year-old Roop Kanwar was burned to death.[85]

Buddhism

The Buddhist view of marriage considers marriage a secular affair and as such, it is not considered a sacrament. Buddhists are expected to follow the civil laws regarding marriage laid out by their respective governments.

Sikhism

In a Sikh marriage, the couple make rounds around the holy book called *Guru Granth Sahib* four times and the holy man speaks some words from the Guru Granth Sahib in the form of kirtan. The ceremony is known as 'Anand Karaj' and represents the holy union of between two souls that are united as one.

Same-sex marriage

A same-sex couple exchanging wedding vows in an Unitarian Universalist Fellowship.

For the most part, religious traditions in the world reserve marriage to heterosexual unions, but there are exceptions including Unitarian Universalist, Metropolitan Community Church, Quaker, United Church of Canada, United Church of Christ and Reform Jewish congregations, some Anglican dioceses, and various Neopagan faiths.[86] [87] This model is currently recognized by various jurisdictions[88] and religious denominations.[89] [90] [91]

Polygyny

Further information: Polygyny and Polygamy

Religious groups have differing views on the legitimacy of polygyny, the practice of a man taking more than one wife. It is allowed in Islam and Confucianism, though in most areas today it is uncommon.[92] [93] Judaism, Christianity and Hinduism have allowed polygny in the past, but it is prohibited by their mainstream modern authorities.[92]

Close-kin marriage

Further information: Cousin marriage and Affinity (canon law)

Religion has commonly weighed in on the matter of which relatives, if any, are allowed to marry. Relations may be by consanguinity or affinity, meaning by blood or by marriage. On the marriage of cousins, Catholic policy has evolved from initial acceptance, through a long period of general prohibition, to the modern-day requirement for a dispensation.[94] Islam has always allowed it, while Hindu strictures vary widely.[95] [96]

Financial considerations

The financial aspects of marriage vary between cultures and have changed over time.

In some cultures, dowries and bride prices continue to be required today. In both cases, the financial arrangements are usually made between the groom (or his family) and the bride's family; with the bride in many cases not being involved in the arrangement, and often not having a choice in whether to participate in the marriage.

In Early Modern Britain, the social status of the couple was supposed to be equal. After the marriage, all the property (called "fortune") and expected inheritances of the wife belonged to the husband.

Dowry

A dowry was not an unconditional gift,[in Early Modern Britain?] but was usually a part of a wider marriage settlement. For example, if the groom had other children, they could not inherit the dowry, which had to go to the bride's children. In the event of her childlessness, the dowry had to be returned to her family, but sometimes not until the groom's death or remarriage.

In some cultures, dowries continue to be required today (for example, in Sudan), while some countries impose restrictions on the payment of dowry. In India, nearly 7,000 women were killed in 2001 over dowries,[97] and activists believe that figures represent only a third of the actual number of such murders.[98]

Bride price and dower

Traditional, formal presentation of the bride price (also known as "sin sot" or "dowry") at an engagement ceremony in Thailand

In other cultures, the groom or his family were expected to pay a bride price to the bride's family for the right to marry the daughter, or dower, which was payable to the bride. This required the groom to work for the bride's family for a set period of time.

In the Jewish tradition, the rabbis in ancient times insisted on the marriage couple entering into a marriage contact, called a *ketubah*. Besides other things, the *ketubah* provided for an amount to be paid by the husband in the event of a divorce or his estate in the event of his death. This amount was a replacement of the biblical dower or bride price, which was payable at the time of the marriage by the groom to the father of the bride. [Exodus 22:15–16] This innovation was put in place because the biblical bride price created a major social problem: many young prospective husbands could not raise the bride price at the time when they would normally be expected to marry. So, to enable these young men to marry, the rabbis, in effect, delayed the time that the amount would be payable, when they would be more likely to have the sum. It may also be noted that both the dower and the *ketubah* amounts served the same purpose: the protection for the wife should her support cease, either by death or divorce. The only difference between the two systems was the timing of the payment. It is the predecessor to the wife's present-day entitlement to maintenance in the event of the breakup of marriage, and family maintenance in the event of the husband not providing adequately for the wife in his will. Another function performed by the *ketubah* amount was to provide a disincentive for the husband contemplating divorcing his wife: he would need to have the amount to be able to pay to the wife.

Morning gifts, which might also be arranged by the bride's father rather than the bride, are given to the bride herself; the name derives from the Germanic tribal custom of giving them the morning after the wedding night. She might have control of this morning gift during the lifetime of her husband, but is entitled to it when widowed. If the amount of her inheritance is settled by law rather than agreement, it may be called dower. Depending on legal systems and the exact arrangement, she may not be entitled to dispose of it after her death, and may lose the property if she remarries. Morning gifts were preserved for many centuries in morganatic marriage, a union where the wife's inferior social status was held to prohibit her children from inheriting a noble's titles or estates. In this case, the morning gift would support the wife and children. Another legal provision for widowhood was jointure, in which property, often land, would be held in joint tenancy, so that it would automatically go to the widow on her husband's death.

Islamic tradition has similar practices. A 'mahr', either immediate or deferred, is the woman's portion of the groom's wealth (divorce) or estate (death). These amounts are usually set on the basis of the groom's own and family wealth and incomes, but in some parts these are set very high so as to provide a disincentive for the groom exercising the divorce, or the husband's family 'inheriting' a large portion of the estate, especially if there are no male offspring from the marriage. In some countries, including Iran, the mahr or alimony can amount to more than a man can ever hope to earn, sometimes up to US$1,000,000 (4000 official Iranian gold coins). If the husband cannot pay the mahr, either in case of a divorce or on demand, according to the current laws in Iran, he will have to pay it by installments. Failure to pay the mahr might even lead to imprisonment.[99]

Modern customs

In many countries today, each marriage partner has the choice of keeping his or her property separate or combining properties. In the latter case, called community property, when the marriage ends by divorce each owns half. In many legal jurisdictions, laws related to property and inheritance provide by default for property to pass upon the death of one party in a marriage firstly to the spouse and secondly to the children. Wills and trusts can make alternative provisions for property succession.

In some legal systems, the partners in a marriage are "jointly liable" for the debts of the marriage. This has a basis in a traditional legal notion called the "Doctrine of Necessities" whereby a husband was responsible to provide necessary things for his wife. Where this is the case, one partner may be sued to collect a debt for which they did not expressly contract. Critics of this practice note that debt collection agencies can abuse this by claiming an unreasonably wide range of debts to be expenses of the marriage. The cost of defense and the burden of proof is then placed on the non-contracting party to prove that the expense is not a debt of the family. The respective maintenance obligations, both during and eventually after a marriage, are regulated in most jurisdictions; alimony is one such method.

Some have attempted to analyze the institution of marriage using economic theory; for example, anarcho-capitalist economist David D. Friedman has written a lengthy and controversial study of marriage as a market transaction (the market for husbands and wives).[100] In the past the economic status of women was enhanced through marriage; however, as more women work nowadays, men gain more economically than women.[101]

Taxation

In some countries, spouses are allowed to average their incomes; this is advantageous to a married couple with disparate incomes. To compensate for this somewhat, many countries provide a *higher* tax bracket for the averaged income of a married couple. While income averaging might still benefit a married couple with a stay-at-home spouse, such averaging would cause a married couple with roughly equal personal incomes to pay more total tax than they would as two single persons. This is commonly called the marriage penalty.

Moreover, when the rates applied by the tax code are not based on averaging the incomes, but rather on the *sum* of individuals' incomes, higher rates will definitely apply to each individual in a two-earner households in progressive tax systems. This is most often the case with high-income taxpayers and is another situation where some consider there to be a marriage penalty.

Conversely, when progressive tax is levied on the individual with no consideration for the partnership, dual-income couples fare much better than single-income couples with similar household incomes. The effect can be increased when the welfare system treats the same income as a shared income thereby denying welfare access to the non-earning spouse. Such systems apply in Australia and Canada, for example.

Other considerations

Sometimes people marry for purely pragmatic reasons, sometimes called a marriage of convenience or sham marriage. For example, according to one publisher of information about "green card" marriages, "Every year over 450,000 United States citizens marry foreign-born individuals and petition for them to obtain a permanent residency (Green Card) in the United States." While this is likely an over-estimate, in 2003 alone 184,741 immigrants were admitted to the U.S. as spouses of U.S. citizens.[102] Many more were admitted as fiancés of US citizens for the purpose of being married within 90 days. Regardless of the number of people entering the US to marry a US citizen, it does not indicate the number of these marriages that are convenience marriages, which number could include some of those with the motive of obtaining permanent residency, but also include many people who are US citizens. One example would be to obtain an inheritance that has a marriage clause. Another example would be to save money on health insurance or to enter a health plan with preexisting conditions offered by the new spouse's employer. Many other situations exist, and, in fact, all marriages have a complex combination of conveniences motivating the parties to marry. A marriage of convenience is one that is devoid of normal reasons to marry.

Some people want to marry a person with higher or lower status than them. Others want to marry people who have similar status. Hypergyny refers to the act of seeking out those who are of slightly higher social status. In most cases, hypergyny refers to women wanting men of higher status. *Isogyny* refers to the act of seeking out those who are of similar status.

Termination

In most societies, the death of one of the partners terminates the marriage, and in monogamous societies this allows the other partner to remarry, though sometimes after a waiting or mourning period.

Many societies also provide for the termination of marriage through divorce. Marriages can also be annulled in some societies, where an authority declares that a marriage never happened. In either event the people concerned are free to remarry (or marry). After divorce, one spouse may have to pay alimony.

The absolute right of two married partners to consent to divorce was only recognized in western nations in recent decades. In the United States no-fault divorce was first recognized in California in 1969 and the final state to recognize it was New York in 1989.[103]

Temporary marriages

Several cultures have practiced temporary and conditional marriages. Examples include the Celtic practice of handfasting and fixed-term marriages in the Muslim community. Pre-Islamic Arabs practiced a form of temporary marriage that carries on today in the practice of Nikah Mut'ah, a fixed-term marriage contract. The prophet Muhammad sanctioned a temporary marriage—sigheh in Iran and muta'a in Iraq— which can provide a legitimizing cover for sex workers.[104] Muslim controversies related to Nikah Mut'ah have resulted in the practice being confined mostly to Shi'ite communities.

Post-marital residence

Early theories explaining the determinants of postmarital residence - with or near the husband's or wife's families, for example - (e.g., Lewis Henry Morgan, Edward Tylor, or George Peter Murdock) connected it with the sexual division of labor. However, to date, cross-cultural tests of this hypothesis using worldwide samples have failed to find any significant relationship between these two variables. However, Korotayev's tests show that the female contribution to subsistence does correlate significantly with matrilocal residence in general; however, this correlation is masked by a general polygyny factor. Although an increase in the female contribution to subsistence tends to lead to matrilocal residence, it also tends simultaneously to lead to general non-sororal polygyny which effectively destroys matrilocality. If this polygyny factor is controlled (e.g., through a multiple regression model), division of labor turns out to be a significant predictor of postmarital residence. Thus, Murdock's hypotheses regarding the relationships between the sexual division of labor and postmarital residence were basically correct, though, as has been shown by Korotayev, the actual relationships between those two groups of variables are more complicated than he expected.[105] [106]

In modern societies we observe a trend toward the neolocal residence.[107]

Contemporary views on marriage

Criticisms

Many people have proposed arguments against marriage for various reasons. These include political and religious criticisms, reference to the divorce rate, as well as celibacy for religious or philosophical reasons.

Controversial views

Many controversies have arisen over the centuries in relation to marriage – including issues relating to the suitability of partners of different denominations, faiths, tribes or races, the acceptable number and minimum age of wives, the rights of partners, especially wives, and wider family obligations. For example, a contemporary controversy of particular significance in the USA concerns the exclusion of homosexual relationships from legal and social recognition and the rights and obligations it provides. Social conservatives opposed to same-sex marriage in some countries claim that any attempt to define marriage to include anything other than the union of one man and one woman would "deprive the term of its fundamental and defining meaning."[108] In other countries, polygamy is a "socially conservative" practice. Advocates of same-faith marriage and same-race marriage may criticize the legalization of interfaith marriage[109] and interracial marriage,[110] respectively.

The state of Massachusetts has sued the U.S. federal government over its definition of marriage. The lawsuit, brought by the first state to legalize gay marriage, said the 1996 Defense of Marriage Act (DOMA) infringed on a state's sovereign right to define marital status. The lawsuit alleges that DOMA infringed on a state's sovereign right to define marital status and is unconstitutional.[111]

Power and gender roles

Feminist theory approaches opposite-sex marriage as an institution traditionally rooted in patriarchy that promotes male superiority and power over women. This power dynamic conceptualizes men as "the provider operating in the public sphere" and women as "the caregivers operating within the private sphere".[112] "Theoretically, women ... [were] defined as the property of their husbands .... The adultery of a woman was always treated with more severity than that of a man."[113] "[F]eminist demands for a wife's control over her own proerty were not met [in parts of Britain] until ... [laws were passed in the late 19th century]."[114] This patriarchal dynamic is contrasted with a conception of egalitarian marriage in which power and labour are divided equally, and not according to gender roles.[112]

The performance of dominant gender roles by men and submissive gender roles by women influence the power dynamic of a marriage.[115] In some American households, women internalize gender role stereotypes and often assimilate into the role of "wife", "mother", and "caretaker" in conformity to societal norms and their male partner. bell hooks states "within the family structure, individuals learn to accept sexist oppression as 'natural' and are primed to support other forms of oppression, including heterosexist domination."[116] "[T]he cultural, economic, political and legal supremacy of the husband" was "[t]raditional ... under English law".[117]

In the US, studies have shown that, despite egalitarian ideals being common, less than half of respondents viewed their opposite-sex relationships as equal in power, with unequal relationships being more commonly dominated by the male partner.[118] Studies also show that married couples find the highest level of satisfaction in egalitarian relationships.[118]

References

[1] Krier, James E.; Gregory S. Alexander, Michael H. Schill, Jesse Dukeminier (2006). *Property*. Aspen Publishers. ISBN 0735557926. Excerpt – page 335: '...at the wedding; hence the importance of including in the marriage ceremony the words, "With all my worldly goods I thee endow."'

[2] Gallagher, Maggie (2002). "What is Marriage For? The Public Purposes of Marriage Law" (http://www.marriagedebate.com/pdf/What is Marriage For.pdf) (PDF). Louisiana law review. . Retrieved 2007-01-08.

[3] Sing Ging Su, (1922) *Chinese Family System* (http://books.google.com/books?id=ZDN-WdK_OoUC&pg=PA55&lpg=PA55&dq=The), p. 54-55. ISBN 0-554-50635-1

[4] Bell, Duran (1997). "Defining Marriage and Legitimacy". *Current Anthropology* **38** (2): 237–254. Bell describes marriage as "a relationship between one or more men (male or female) in severalty to one or more women that provides those men with a demand-right of sexual access within a domestic group and identifies women who bear the obligation of yielding to the demands of those specific men."

[5] Westermarck, Edvard, (1921) *The History of Human Marriage Volume 1* (http://books.google.com/books?id=ShU3N7_GaMYC), p. 71. ISBN 0-7661-4618-9.

[6] Westermarck, Edvard, (1936) *The Future of Marriage in Western Civilization* (http://books.google.ca/books?id=gZ8LyGDtSZIC), p. 3.

[7] *Notes and Queries on Anthropology*. Royal Anthropological Institute. 1951. pp. , p. 110.

[8] Gough, E. Kathleen (1959). "The Nayars and the Definition of Marriage". *Royal Anthropological Institute of Great Britain and Ireland*: 89:23–34 Nuer female-female marriage is done to keep property within a family that has no sons. It is not a form of lesbianism.

[9] Gudeman, Stephen, (1976) *Relationships, residence and the individual*, p. 131.

[10] Leach, Edmund (1955). "Polyandry, Inheritance and the Definition of Marriage". *Man*.

[11] Oxford English Dictionary 11th Edition, "marriage"

[12] Hobhouse, Leonard Trelawny (1906) Morals in evolution: a study in comparative ethics (Page 180)

[13] Hanlon & White, p. 120.

[14] Hanlon & White, p. 119.

[15] Hanlon & White, p. 116.

[16] *Group Marriage* (http://www.britannica.com/EBchecked/topic/247131/group-marriage). Encyclopædia Britannica Online. . Retrieved January 28, 2010

[17] "Marriage, a History." *Psychology Today*, May 01, 2005

[18] "Magnus Hirschfeld Archive of Sexology" (http://www2.hu-berlin.de/sexology/ATLAS_EN/html/history_of_marriage_in_western.html). .hu-berlin.de. . Retrieved 2010-08-27.

[19] "Roman empire.net marriage" (http://www.roman-empire.net/society/soc-marriage.html). Roman-empire.net. . Retrieved 2010-08-27.

[20] Treggiari, Susan. Roman Marriage: Iusti Coniuges from the Time of Cicero to the Time of Ulpian (http://books.google.com/books?id=I0J1A6o4GuQC&lpg=PP1&dq=susan treggiari&pg=PA39#v=onepage&q=&f=false). p. 39.

[21] "St. Ignatius of Antioch to Polycarp (Roberts-Donaldson translation)" (http://www.earlychristianwritings.com/text/ignatius-polycarp-longer.html). Earlychristianwritings.com. 2006-02-02. . Retrieved 2010-08-27.

[22] Pernoud, Régine (2000). *Those terrible Middle Ages: debunking the myths*. San Francisco: Ignatius Press. p. 102. ISBN 978-0-89870-781-6.

[23] upenn.edu Excerpt from Marriage, Sex, and Civic Culture in Late Medieval London (http://www.upenn.edu/pennpress/book/toc/14042.html) *"the sacramental bond of marriage could be made only through the freely given consent of both parties"*

[24] "marriage.about.com" (http://marriage.about.com/cs/generalhistory/a/marriagehistory.htm). marriage.about.com. 2010-06-16. . Retrieved 2010-08-27.

[25] "Marriage Records" (http://www.exploregenealogy.co.uk/FindingEarlyMarriageRecords.html). Exploregenealogy.co.uk. 2007-10-29. . Retrieved 2010-08-27.

[26] Schofield, Phillipp R. 2003. Peasant and community in Medieval England, 1200–1500. Medieval culture and society. New York: Palgrave-Macmillan. p 98.

[27] "History of Marriage". *Catholic Encyclopedia*. New York: Robert Appleton Company. 1913.

[28] *West's Encyclopedia of American Law, 2nd Edition*. Thomson Gale, 2005. ISBN 0-7876-6367-0

[29] Witte Jr., John (1997). *From Sacrament to Contract: Marriage, Religion, and Law in the Western Tradition*. Westminster John Knox Press. pp. 39–40. ISBN 0664255434.

[30] Leneman, Leah (1999). "The Scottish Case That Led to Hardwicke's Marriage Act" (http://www.historycooperative.org/journals/lhr/17.1/leneman.html). Law and History Review. .

[31] Gillis, John R. (1985). *For Better, for Worse: British Marriages, 1600 to the Present* (http://books.google.com/?id=t3kiLAQxrnMC). Oxford University Press. p. 92. ISBN 019503614X. .

[32] "Marriage|LII / Legal Information Institute" (http://topics.law.cornell.edu/wex/Marriage). Topics.law.cornell.edu. 2010-08-19. . Retrieved 2010-08-27.

[33] Westermarck, Edward Alexander (1903). *The History of Human Marriage*. Macmillan and Co., Ltd., London. ISBN 1402185480 (reprint).

[34] Neill, James (2009). *The origins and role of same-sex... – Google Books* (http://books.google.com/books?id=1ha9GgWNmy0C&pg=PT267). ISBN 9780786435135. . Retrieved 2010-08-27.

[35] Herdt, Gilbert H (1993). *Ritualized Homosexuality in Melanesia – Google Books* (http://books.google.com/?id=mlFp0nFhvbwC&dq=Ritualized+Homosexuality+Herdt&printsec=frontcover#v=onepage&q=&f=false). ISBN 9780520080966. . Retrieved 2010-08-27.

[36] Alderson, Kevin; Lahey, Kathleen A. (2004). *Same-Sex Marriage: The Personal and the Political* (http://books.google.com/?id=hR0_CoNj6GAC&pg=RA1-PA16). Insomniac Press. p. 16. ISBN 978-1-894663-63-2. .

[37] Hinsch, Bret (1990). *Passions of the Cut Sleeve: The Male Homosexual Tradition in China*. Reed Business Information, Inc.. ISBN 0520078691.

[38] *ubi scelus est id, quod non proficit scire, ubi venus mutatur in alteram formam, ubi amor quaeritur nec videtur, iubemus insurgere leges, armari iura gladio ultore, ut exquisitis poenis subdantur infames, qui sunt vel qui futuri sunt rei.* Ancientrome.ru (http://ancientrome.ru/ius/library/codex/theod/liber09.htm#7) "where that crime is found, which is unfit even to know, we command the law to arise armed with an avenging sword that the infamous men who are, or shall in future be guilty of it, may undergo the most severe punishments." *translation by Lord Blackstone,* Commentaries on the Laws of England *(Oxford: Clarendon Press, 1769, Vol. IV, pp. 215–216.*

[39] Kuefler, Mathew (2007). "The Marriage Revolution in Late Antiquity: The Theodosian Code and Later Roman Marriage Law" (http://jfh.sagepub.com/cgi/content/short/32/4/343). *Journal of Family History* **32** (4): 343–370. doi:10.1177/0363199007304424. .

[40] Murdock, 1949, p. 24. "*group marriage* or a marital union embracing at once several men and several women."

[41] Child Marriage Factsheet: State of World Population 2005 – UNFPA (http://www.unfpa.org/swp/2005/presskit/factsheets/facts_child_marriage.htm)

[42] Turkish Civil and Penal Code Reforms from a Gender Perspective: The Success of two Nationwide Campaigns (http://www.wwhr.org/images/CivilandPenalCodeReforms.pdf)PDF (6.21 MB) (p. 18)

[43] Rosenblatt, Paul C. (2006). *Two in a Bed: The Social System of Couple Bed Sharing* (http://www.sunypress.edu/details.asp?id=61306). State University of New York Press. ISBN 0-7914-6829-1. .

[44] Sanday, Peggy Reeves (2002). *Women at the center: life in a modern matriarchy*. Cornell University Press. ISBN 0-8014-8906-7.

[45] Gargan, Edward A. (2001-03-19). "China's New Brides Put Freedom First/All perks, no work in 'walking marriages'". Newsday. pp. A.04.

[46] Karam, Souhail (July 21, 2006). "Misyar offers marriage-lite in strict Saudi society" (http://www.boston.com/news/world/middleeast/articles/2006/07/21/misyar_offers_marriage_lite_in_strict_saudi_society/). Reuters. .

[47] Income Tax Assessment Act 1997 – Sect 995.1(1) (http://www.austlii.edu.au/au/legis/cth/consol_act/itaa1997240/s995.1.html): ""spouse" of an individual includes: (a) another individual (whether of the same sex or a different sex) with whom the individual is in a relationship that is registered under a * State law or * Territory law prescribed for the purposes of section 22B of the Acts Interpretation Act 1901 as a kind of relationship prescribed for the purposes of that section; and (b) another individual who, although not legally married to the individual, lives with the individual on a genuine domestic basis in a relationship as a couple."

[48] Ontario Consultants on Religious Tolerance (2006-12-31). "Human sexuality and gender topics: Subjects of major concern to many faith groups" (http://www.religioustolerance.org/chr_sex.htm). Religioustolerance.org. . Retrieved 2007-02-04.

[49] Jones, Richard E.; Kristin H. Lopez (2006). *Human Reproductive Biology, Third Edition*. Academic Press. ISBN 0120884658.

[50] Ventura, SJ. (1995) (PDF). *Births to unmarried mothers: United States, 1980–92.* (http://www.cdc.gov/nchs/data/series/sr_21/sr21_053.pdf). National Center for Health Statistics. ISBN 0-8406-0507-2. . Retrieved 2008-12-29.

[51] Harris, Gardiner (2007-12-06). "Teenage Birth Rate Rises for First Time Since '91" (http://www.nytimes.com/2007/12/06/washington/06birth.html?em&ex=1197176400&en=62f9e9412af61f8b&ei=5087). New York Times. . Retrieved 2010-04-10.

[52] Bawah, AA.; Akweongo P, Simmons R, Phillips JF. (1999). "Women's fears and men's anxieties: the impact of family planning on gender relations in northern Ghana" (http://www.popcouncil.org/pdfs/councilarticles/sfp/SFP301Bawah.pdf) (PDF). *Studies in Family Planning* (Population Council) **30** (1): 54–66. doi:10.1111/j.1728-4465.1999.00054.x. PMID 10216896. . Retrieved 2008-12-29.

[53] "Common Law Marriage FAQ." *Nolo.* July 31, 2009. Nolo.com (http://www.nolo.com/article.cfm/objectID/709FAEE4-ABEA-4E17-BA34836388313A3C/118/304/192/FAQ/)

[54] Leach, Edmund (1968). Paul Bonannan and John Middleton. ed. *Marriage, Family, and Residence*. The Natural History Press. ISBN 1121644708.

[55] "Adultery" (http://www.britannica.com/EBchecked/topic/6618/adultery) *Encyclopedia Brittanica*

[56] Smith, Anna Marie. "Sex Scandals, 'Responsible Fatherhood' and the 2008 Election Campaign: When 'Sex Talk' Trumps Race and Class" (http://www.barnard.edu/sfonline/sexecon/smith_05.htm). p. 5. .

[57] Perez, Evan. "DOJ Shifts Tone on Marriage Act." *The Wall Street Journal,* U.S. ed., August 18, 2009.

[58] "Same-Sex Marriage, Civil Unions and Domestic Partnerships." National Conference of State Legislatures. August 18, 2009:

[59] Conniff, Richard (2003-08-01). "Richard Conniff. "Go Ahead, Kiss Your Cousin."" (http://discovermagazine.com/2003/aug/featkiss). Discovermagazine.com. . Retrieved 2010-08-27.

[60] Kershaw, Sarah (November 26, 2009). "Shaking Off the Shame" (http://www.nytimes.com/2009/11/26/garden/26cousins.html?pagewanted=1&_r=1). *The New York Times*. .

[61] See Article 809 of the Korean Civil Code and *The first ten years of the Korean Constitutional Court* (http://www.ccourt.go.kr/home/english/download/decision_10years.pdf). Constitutional Court of Korea. p. 242 (p.256 of the PDF).

[62] *This article incorporates text from the 1901–1906 Jewish Encyclopedia article "marriage"* (http://www.jewishencyclopedia.com/view.jsp?letter=M&artid=213)*, a publication now in the public domain.*

[63] *This article incorporates text from the 1903 Encyclopaedia Biblica article "MARRIAGE", a publication now in the public domain.*

[64] William Robertson Smith, *Kinship and Marriage in early Arabia*, (1885), 81

[65] Genesis 29:9; Exodus 2:16;, 8:13

[66] "Why Marry?" (http://www.chabad.org/library/article_cdo/aid/448425/jewish/Why-Marry.htm). Chabad.org. . Retrieved 2007-12-19.

[67] "Religions – Christianity: Marriage and weddings" (http://www.bbc.co.uk/religion/religions/christianity/ritesrituals/weddings_1.shtml). BBC. . Retrieved 2010-08-27.

[68] Batey, Richard (1971). *New Testament nuptial imagery* (http://books.google.com/?id=Zz4VAAAAIAAJ&pg=PA1). Brill Archive. p. 1. .

[69] See also [Mark 10:7], Gen. 2:24, Matt. 19:5, Eph. 5:31

[70] "Sacrament of Marriage". *Catholic Encyclopedia*. New York: Robert Appleton Company. 1913.

[71] "GCSE Bitesize: Marriage" (http://www.bbc.co.uk/schools/gcsebitesize/rs/relationships/chmarriageanddivorcerev1.shtml). BBC. . Retrieved 2010-08-27.

[72] "Catechism of the Catholic Church, Second Edition, Article Seven, Paragraph 1612" (http://www.vatican.va/archive/catechism/p2s2c3a7.htm#I). Vatican.va. . Retrieved 2010-08-27.

[73] "Catechism of the Catholic Church, Second Edition, Article Seven, Paragraph 1601" (http://www.vatican.va/archive/catechism/p2s2c3a7.htm#I). Vatican.va. . Retrieved 2010-08-27.

[74] Sacrament of Holy Matrimony, by P. McLachlan Catholic-pages.com (http://www.catholic-pages.com/marriage/sacrament.asp)

[75] 1983 CODE c.1056 (Latin-English edition of the Code of Canon Law and English-language translation of the 5th Spanish-language edition of the commentary prepared under the responsibility of the Instituto Martin de Azpilcueta, 1993)

[76] "Catechism of the Catholic Church, Second Edition, Article Seven, Paragraph 1665" (http://www.vatican.va/archive/catechism/p2s2c3a7.htm#brief). Vatican.va. . Retrieved 2010-08-27.

[77] "The Family: A Proclamation to the World" (http://www.lds.org/library/display/0,4945,161-1-11-1,00.html). Lds.org. 1995-09-23. . Retrieved 2010-08-27.

[78] The method of pronouncing the marriage formula (http://www.sistani.org/local.php?modules=nav&nid=2&bid=59&pid=3079)

[79] Marriage formula (http://www.sistani.org/local.php?modules=nav&nid=2&bid=59&pid=3078)

[80] Conditions of pronouncing Nikah (http://www.sistani.org/local.php?modules=nav&nid=2&bid=59&pid=3080)

[81] Women with whom matrimony is Haraam (http://www.sistani.org/local.php?modules=nav&nid=2&bid=59&pid=3083)

[82] Smith, Peter (2000). "Marriage". *A concise encyclopedia of the Bahá'í Faith*. Oxford: Oneworld Publications. pp. 232–233. ISBN 1-85168-184-1.

[83] " Shunned from society, widows flock to city to die (http://www.cnn.com/2007/WORLD/asiapcf/07/05/damon.india.widows/index.html)". CNN.com. July 5, 2007.

[84] "Widow Remarriage under Hindu Laws | Indian Laws" (http://www.lawisgreek.com/widow-remarriage-under-hindu-laws/). Lawisgreek.com. . Retrieved 2010-08-27.

[85] "The New York Times, 1987" (http://query.nytimes.com/gst/fullpage.html?res=9B0DE6D61139F933A1575AC0A961948260). 1987-09-20. . Retrieved 2008-05-31.

[86] "World Religions and Same-Sex Marriage" (http://marriagelaw.cua.edu/publications/wrr.pdf), Marriage Law Project, Columbus School of Law at The Catholic University of America, Washington, D.C., July 2002 revision

[87] "Affirming Congregations, The Episcopal Church and Ministries of the United Church of Canada" (http://www.united-church.ca/exploring/marriage/affirmingcongregations). United-church.ca. . Retrieved 2010-08-27.

[88] Arce, Rose. *Massachusetts court upholds same-sex marriage*. Feb. 6, 2004. CNN (http://www.cnn.com/2004/LAW/02/04/gay.marriage/). Retrieved Feb. 17, 2007.

[89] "Religious Groups' Official Positions on Same-Sex Marriage" (http://pewforum.org/docs/?DocID=291). pewforums.org. . Retrieved 2008-10-16.

[90] Shaila Dewan (July 5, 2005). "United Church of Christ Backs Same-Sex Marriage" (http://www.nytimes.com/2005/07/05/national/05church.html). New York Times. . Retrieved 2008-10-16.

[91] "Unitarian Universalist Association of Congregations BLGT Community guide" (http://www.uua.org/visitors/justicediversity/6252.shtml). . Retrieved 2008-10-16.

[92] "The Dinah Project" (http://www.dinahproject.com/articles_view_details.asp?id=217). The Dinah Project. . Retrieved 2010-08-27.

[93] "Romance of Three Kingdoms – by Luo Guanzhong – History" (http://threekingdoms.com/history.htm). Threekingdoms.com. . Retrieved 2010-08-27.

[94] Ottenheimer, Martin (1996). "Chapter 3". *Forbidden Relatives: The American Myth of Cousin Marriage*. University of Illinois.

[95] "Islamic View on Marrying Cousins – IslamonLine.net – Ask The Scholar" (http://www.islamonline.net/servlet/Satellite?pagename=IslamOnline-English-Ask_Scholar/FatwaE/FatwaE&cid=1119503544772). IslamonLine.net. . Retrieved 2010-08-27.

[96] Srinivas, Mysore Narasimhachar (1980). *India: Social Structure*. Delhi: Hindustan Publishing Corporation. p. 55.

[97] " India's dowry deaths (http://news.bbc.co.uk/2/hi/programmes/crossing_continents/3071963.stm)". BBC News. July 16, 2003.

[98] " Women: killed by greed and oppression (http://www.time.com/time/international/1995/950911/women.india.html)". TIME Magazine. September 11, 1995 Volume 146, No. 11

[99] "A translation of some parts of the Civil Code of Iran" (http://www.international-divorce.com/iran_divorce.htm). International-divorce.com. . Retrieved 2010-08-27.

[100] "The Economics of Love and Marriage" (http://www.daviddfriedman.com/Academic/Price_Theory/PThy_Chapter_21/PThy_Chap_21.html). Daviddfriedman.com. . Retrieved 2010-08-27.

[101] Richard Fry, D'Vera Cohn: *Women, Men and the New Economics of Marriage* (http://pewsocialtrends.org/pubs/750/new-economics-of-marriage), Pew Research Center, January 19, 2010

[102] Immigration to the United States: Fiscal years 1820–2003 (http://www.dhs.gov/xlibrary/assets/statistics/yearbook/2003/2003IMMtables.pdf)PDF (2.03 MB)

[103] "No-Fault Divorce – The Pros and Cons Of No-Fault Divorce" (http://divorcesupport.about.com/od/maritalproblems/i/nofault_fault_2.htm). Divorcesupport.about.com. 2010-07-30. . Retrieved 2010-08-27.

[104] İlkkaracan, Pınar (2008). *Deconstructing sexuality in the Middle East: challenges and discourses* (http://books.google.com/books?id=pnGwP9-FhxYC&pg=PA36&dq#v=onepage&q=&f=false). Ashgate Publishing, Ltd.. p. 36. ISBN 0754672352. .

[105] Korotayev A. Form of marriage, sexual division of labor, and postmarital residence in cross-cultural perspective: A reconsideration. Journal of anthropological research ISSN 0091-7710. 2003, Vol. 59, No. 1, pp. 69–89 (http://cat.inist.fr/?aModele=afficheN&cpsidt=14683590)

[106] Korotayev A. Division of Labor by Gender and Postmarital Residence in Cross-Cultural Perspective: A Reconsideration (http://cat.inist.fr/?aModele=afficheN&cpsidt=14683590). Cross-Cultural Research. 2003, Vol. 37, No. 4, pp.335–372 (DOI: 10.1177/1069397103253685)

[107] Marriage, Family, and Kinship: Comparative Studies of Social Organization, by Melvin Ember and Carol R. Ember. New Haven: HRAF Press, 1983)

[108] "Same-Sex Marriage" (http://www.rabbis.org/news/article.cfm?id=100556) (Press release). Rabbinical Council of America. 2004-03-30. .

[109] 2 Corinthians 6:14 *Do not be yoked together with unbelievers. For what do righteousness and wickedness have in common? Or what fellowship can light have with darkness?*

[110] Werner Sollors (2000-10-19). Interracialism: Black-White Intermarriage in American History, Literature, and Law (Sollors, Werner ed.). Oxford: Oxford University Press. ISBN 0-19-512856-7

[111] Shishkin, Philip. "Massachusetts Sues U.S. Over Definition of Marriage." *Wall Street Journal – Eastern Edition*; 7/9/2009, Vol. 254 Issue 7, pA3, 0p, 1 bw

[112] Briana Weadock, "Disciplining Marriage: Gender, Power, and Resistance", (http://www.allacademic.com/meta/p110303_index.html), 2000

[113] Evans, Tanya, *Women, Marriage and the Family*, in Barker, Hannah, & Elaine Chalus, eds., *Women's History: Britain, 1700–1850: An Introduction* (Oxon/London: Routledge, 2005 (ISBN 0-415-29177-1)), p. 64 (author Evans postdoctoral research fellow, Ctr. for Contemp. Brit. Hist., Institute for Historical Research, London, editor Barker sr. lecturer history, Univ. of Manchester, & editor Chalus sr. lecturer history, Bath Spa Univ. Coll.

[114] Evans, Tanya, *Women, Marriage and the Family, op. cit.*, in Barker, Hannah, & Elaine Chalus, eds., *Women's History, op. cit.*, p. 66 & n. 69.

[115] Veronica Jaris Tichenor, "Thinking About Gender and Power in Marriage", The Kaleidoscope of Gender, 2010

[116] bell hooks, Feminist Theory From Margin to Center, 2000

[117] Barnett, Hilaire A, *Introduction to Feminist Jurisprudence* (London, G.B., U.K.: Cavendish Publishing, 1998 (ISBN 1 85941 237 8)), p. 35 and, per p. 35 n. 35, see chap. 3 (author then of Queen Mary & Westfield Coll., Univ. of London).

[118] Sprecher, Susan; Felmlee, Diane, "The Balance of Power in Romantic Heterosexual Couples Over Time from 'His' and 'Her' Perspectives", (http://business.highbeam.com/435388/article-1G1-20247079/balance-power-romantic-heterosexual-couples-over-time), 1997

External links

- African Marriage Rituals (http://www.africanholocaust.net/news_ah/africanmarriageritual.html)
- *For Better, for Worse: British Marriages, 1600 to the Present* (http://books.google.com/books?id=t3kiLAQxrnMC) John Gillis. 1985. Oxford University Press. ISBN 0-19-503614-X
- 'Forever and a Day' or 'Just One Night'? (http://www.in-mind.org/issue-4/forever-and-a-day-or-just-one-night.html) On Adaptive Functions of Long-Term and Short-Term Romantic Relationships
- "Legal Regulation of Marital Relations: An Historical and Comparative Approach – Gautier 19 (1): 47 – International Journal of Law, Policy and the Family" (http://lawfam.oxfordjournals.org/cgi/content/abstract/19/1/47).
- "Marriage – its various forms and the role of the State" (http://www.bbc.co.uk/radio4/history/inourtime/inourtime_20020321.shtml) on BBC Radio 4's *In Our Time* featuring Janet Soskice, Frederik Pedersen and Christina Hardyment
- Radical principles and the legal institution of marriage: domestic relations law and social democracy in Sweden (http://lawfam.oxfordjournals.org/cgi/content/abstract/4/2/154) – Bradley 4 (2): 154 – International Journal of Law, Policy and the Family
- Recordings (http://www.thehist.com/index.php?option=com_docman&task=cat_view&gid=145&Itemid=673) & Photos (http://www.thehist.com/index.php?option=com_content&task=view&id=709&Itemid=672) from a College Historical Society debate on the role of marriage in modern life, featuring Senator

David Norris and Senator Ronan Mullen.

- The Delights of Wisdom Concerning Marriage ("Conjugial") Love, After Which follows the Pleasures of Insanity Concerning Scortatory Love. (http://swedenborgdigitallibrary.org/contets/MLtc.html) by Emanuel Swedenborg (Swedenborg Society 1953)

rue:Манжелство

Poverty threshold

The **poverty threshold**, or **poverty line**, is the minimum level of income deemed necessary to achieve an adequate standard of living in a given country.[1] In practice, like the definition of poverty, the official or common understanding of the poverty line is significantly higher in developed countries than in developing countries.[2] [3]

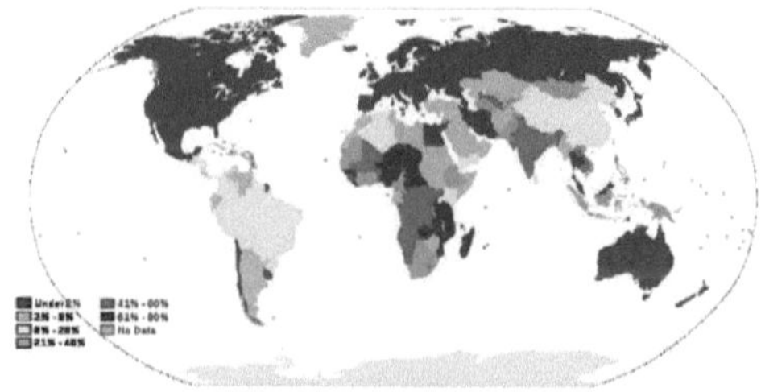

Percentage of population living on less than $1.25 per day. UN estimates 2000-2007.

The common **international poverty line** has in the past been roughly $1 a day.[4] In 2008, the World Bank came out with a revised figure of $1.25 at 2005 purchasing-power parity (PPP).[5]

Determining the poverty line is usually done by finding the total cost of all the essential resources that an average human adult consumes in one year.[6] This approach is needs-based in that an assessment is made of the minimum expenditure needed to maintain a tolerable life. This was the original basis of the poverty line in the United States, whose calculation was simplified to be based solely on the cost of food and is updated each year. The largest of these expenses is typically the rent required to live in an apartment, so historically, economists have paid particular attention to the real estate market and housing prices as a strong poverty line affector.

Individual factors are often used to account for various circumstances, such as whether one is a parent, elderly, a child, married, etc. The poverty threshold may be adjusted each year.

The poverty threshold is useful as an economic tool with which to measure such people and consider socioeconomic reforms such as welfare and unemployment insurance to reduce poverty.

Absolute poverty

A measure of *absolute poverty* quantifies the number of people below a fixed real poverty threshold. It is a level of poverty as defined in terms of the minimal requirements necessary to afford minimal standards of food, clothing, health care and shelter.[7] (See definition from Glossary of Sociology.) For the measure to be absolute, the line must be the same in different countries, cultures, and technological levels. Such an absolute measure should look only at the individual's power to consume and it should be independent of any changes in income distribution. Such a measure is possible only when all consumed goods and services are counted and when PPP-exchange rates are used (see purchasing power parity). The intuition behind an absolute measure is that mere survival takes essentially the same amount of resources across the world and that everybody should be subject to the same standards if meaningful comparisons of policies and progress are to be made. Notice that if everyone's real income in an economy increases, and the income distribution does not change, absolute poverty will decline.

Measuring poverty by an absolute threshold has the advantage of applying the same standard across different locations and time periods, making comparisons easier. On the other hand, it suffers from the disadvantage that any absolute poverty threshold is to some extent arbitrary; the amount of wealth required for survival is not the same in all places and time periods. For example, a person living in far northern Scandinavia requires a source of heat during colder months, while a person living on a tropical island does not.

This type of measure is often contrasted with measures of relative poverty (see below), which classify individuals or families as "poor" not by comparing them to a fixed cutoff point, but by comparing them to others in the population under study.

The term *absolute poverty* is also sometimes used as a synonym for extreme poverty. Absolute poverty is the absence of enough resources (such as money) to secure basic life necessities.

According to a UN declaration that resulted from the World Summit on Social Development in Copenhagen in 1995, absolute poverty is "a condition characterised by severe deprivation of basic human needs, including food, safe drinking water, sanitation facilities, health, shelter, education and information. It depends not only on income but also on access to services."[8]

David Gordon's paper, "Indicators of Poverty & Hunger", for the United Nations, further defines absolute poverty as the absence of any two of the following eight basic needs:[8]

- *Food:* Body Mass Index must be above 16.
- *Safe drinking water:* Water must not come from solely rivers and ponds, and must be available nearby (less than 15 minutes' walk each way).
- *Sanitation facilities:* Toilets or latrines must be accessible in or near the home.
- *Health:* Treatment must be received for serious illnesses and pregnancy.
- *Shelter:* Homes must have fewer than four people living in each room. Floors must not be made of dirt, mud, or clay.
- *Education:* Everyone must attend school or otherwise learn to read.
- *Information:* Everyone must have access to newspapers, radios, televisions, computers, or telephones at home.
- *Access to services:* This item is undefined by Gordon, but normally is used to indicate the complete panoply of education, health, legal, social, and financial (credit) services.

For example, a person who lives in a home with a mud floor is considered severely deprived of shelter. A person who never attended school and cannot read is considered severely deprived of education. A person who has no newspaper, radio, television, or telephone is considered severely deprived of information. All people who meet any two of these conditions — for example, they live in homes with mud floors *and* cannot read — are considered to be living in absolute poverty.

Relative poverty

A measure of *relative poverty* defines "poverty" as being below some relative poverty threshold. For example, the statement that "households with an accumulated income less than 60% of the median equivalized household disposable income are living in poverty" uses a relative measure to define poverty. In this system, if everyone's real income in an economy increases, but the income distribution stays the same, then the rate of relative poverty will also stay the same.

Relative poverty measurements can sometimes produce odd results, especially in small populations. For example, if the median household in a wealthy neighborhood earns US$1 million each year, then a family that earns US$100,000 would be considered poor on the relative poverty scale, even though such a family could meet all of its basic needs and much more. At the other end of the scale, if the median household in a very poor neighborhood earned only 50% of what it needs to buy food, then a person who earned the median income would not be considered poor on a relative poverty scale, even though the person is clearly poor on an absolute poverty scale.

Measures of relative poverty are almost the same as measuring income inequality: If a society gets a more equal income distribution, relative poverty will fall. Following this, some argue that the term *relative poverty* is itself misleading and that *income inequality* should be used instead. They point out that if society changed in a way that hurt high earners more than low ones, then relative poverty would decrease, but every citizen of the society would be worse off. Likewise in the reverse direction: it is possible to reduce absolute poverty while increasing relative poverty.

The term *relative poverty* can also be used in a different sense to mean "moderate poverty" — for example, a standard of living or level of income that is high enough to satisfy basic needs (like water, food, clothing, shelter, and basic health care), but still significantly lower than that of the majority of the population under consideration.

Basic needs

Some measurements combine certain aspects of absolute and relative measures. For example, the Fraser Institute publishes a basic needs poverty measure for Canada. According to the Fraser Institute, "the basic-needs approach is partly absolute (the list [of necessities] is limited to items required for long-term physical well-being) and partly relative, reflecting the standards that apply in the individual's own society at the present time."[9] The Fraser Institute's list of necessities for living creditably in Canada includes not only food, shelter, clothing, and health care, but also personal care, furniture, transportation, communication, laundry, and home insurance. It is a criticized for not including any entertainment items like cable television, daily newspapers, and tickets to movies or sporting events.[9] [10]

National poverty lines

National estimates are based on population-weighted subgroup estimates from household surveys. Definitions of the poverty line may vary considerably among nations. For example, rich nations generally employ more generous standards of poverty than poor nations. Thus, the numbers are not comparable among countries.

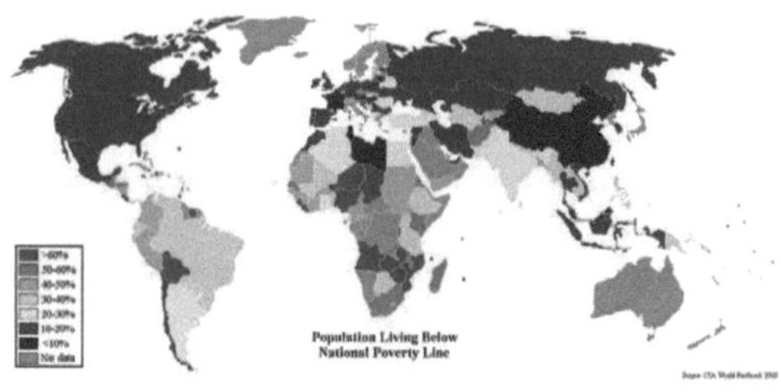

CIA World Factbook based map showing the percentage of population by country living below that country's official poverty line.

In 2009, in the United States of America, the poverty threshold for a single person under 65 was US$11,161; the threshold for a family group of four, including two children, was US$21,756.[11] [12] According to the U.S. Census Bureau data released Tuesday September 13th, 2011, the nation's poverty rate rose to 15.1% in 2010.

In the UK, "more than five million people – over a fifth (23 percent) of all employees – were paid less than £6.67 an hour in April 2006. This is based on a low pay rate of 60 percent of full-time median earnings, equivalent to a little over £12,000 a year for a 35-hour working week. In April 2006, a 35-hour week would have earned someone £9,191 a year – before tax or National Insurance".[13] [14]

India's official poverty level, on the other hand, is split according to rural vs. urban thresholds. For urban dwellers, the poverty line is defined as living on less than 538.60 rupees (approximately USD $12) per month, whereas for rural dwellers, it is defined as living on less than 356.35 rupees per month (approximately USD $7.50).[15] By this measure, only 27.5% of Indians live in poverty, whereas by the World Bank standard of $1.25 per day, 42% of Indians live in poverty – this is the third highest rate in South Asia after Bangladesh and Bhutan.

In the year 2011 there was a crossfire between National Advisory Council of India (NAC) & Planning Commission of India. The row was sparked when Planning Commission informed the Supreme Court that for an average Indian urban dweller, Rs.32 per day is adequate to carry out his/her living and that those who earn more than Rs.32 ($ 2 in terms of PPP) in Urban areas & Rs.25 ($ 1.56 in terms of PPP) in Rural areas shall not be considered as poor anymore. Former NAC member Jean Dreze said the health expenditure estimated by the Tendulkar panel was "barely enough to buy an aspirin". Food rights activist Colin Gonsalves described the definition as "shocking". The per capita expenditure of Rs 32 per day in cities won't be sufficient even for food. He also quoted saying that "A person cannot feed himself properly with this money. Where from he will spend on clothing, health and other things?".

Criticisms

Using a poverty threshold is problematic because having an income marginally above it is not substantially different from having an income marginally below it: the negative effects of poverty tend to be continuous rather than discrete, and the same low income affects different people in different ways. To overcome this problem, poverty indices are sometimes used instead; see income inequality metrics.

A poverty threshold relies on a quantitative, or purely numbers-based, measure of income. If other human development-indicators like health and education are used, they must be quantified, which is not a simple (if even achievable) task.

Overstating poverty

In-kind gifts, whether from public or private sources, are not counted when calculating a poverty threshold. For example, if a parent pays the rent on an apartment for an adult son directly to the apartment owner, instead of giving the money to the son to pay the rent, then that money does not count as income to the son. If a church or non-profit organization gives food to an elderly person, the value of the food is not counted as income to the elderly person. Rea Hederman, a senior policy analyst in the Center for Data Analysis at the Heritage Foundation, said

> The official poverty measure counts only monetary income. It considers antipoverty programs such as food stamps, housing assistance, the Earned Income Tax Credit, Medicaid and school lunches, among others, "in-kind benefits" – and hence not income. So, despite everything these programs do to relieve poverty, they aren't counted as income when Washington measures the *poverty rate.*[16]

Studies measuring the difference between income before and after taxes and government transfers, however, have found that without these programs poverty would be roughly 30% to 40% higher than the official poverty line indicates, despite many of their benefits not being counted as income.[17] [18]

See also

- Asset poverty
- Income deficit
- Measuring poverty
- List of countries by percentage of population living in poverty
- Social issues in Brazil
- Poverty in China
- Poverty in India
- Poverty in the United Kingdom
- Poverty in the United States

References

[1] Ravallion, Martin *Poverty Comparisons: A Guide to Concepts and Methods*. Living Standards Measurement Papers, The World Bank, 1992, p. 25

[2] Hagenaars, Aldi & de Vos, Klaas *The Definition and Measurement of Poverty*. Journal of Human Resources, 1988

[3] Hagenaars, Aldi & van Praag, Bernard *A Synthesis of Poverty Line Definitions*. Review of Income and Wealth, 1985

[4] Sachs, Jeffrey D. *The End of Poverty* 2005, p. 20

[5] Ravallion, Martin; Chen Shaohua & Sangraula, Prem *Dollar a day* The World Bank Economic Review, 23, 2, 2009, pp. 163-184

[6] {Poverty Lines-Martin Ravallion, in The New Palgrave Dictionary of Economics, 2nd Edition, London: Palgrave Macmillan}

[7] http://dictionary.babylon.com/absolute%20poverty/

[8] "Indicators of Poverty and Hunger" (http://www.un.org/esa/socdev/unyin/documents/ydiDavidGordon_poverty.pdf). . Retrieved 2008-02-14.

[9] Poverty in Canada: 2006 Update (http://www.fraserinstitute.org/COMMERCE.WEB/product_files/PovertyinCanada2006.pdf), Fraser Institute, November 2006, URL accessed 14 February 2008

[10] "Defining and Re-Defining Poverty: A CCSD Perspective" (http://www.ccsd.ca/pubs/2001/povertypp.htm), October 2001, Canadian Council On Social Development

[11] "Poverty Thresholds 2009" (http://www.census.gov/hhes/www/poverty/about/overview/measure.html). . Retrieved 2010-12-22.

[12] US Census Bureau. "How the Census Bureau Measures Poverty" (http://www.census.gov/hhes/www/poverty/data/threshld/thresh09.xls). . Retrieved 22 December 2010.

[13] Working out of Poverty: A study of the low paid and the working poor by Graeme Cooke and Kayte Lawton

[14] IPPR Article: "Government must rescue 'forgotten million children' in poverty" (http://www.ippr.org.uk/pressreleases/?id=2965)

[15] "POVERTY ESTIMATES FOR 2004-05" (http://planningcommission.gov.in/news/prmar07.pdf). . Retrieved 2009-11-19.

[16] "Poor poverty yardsticks" (http://www.heritage.org/Press/Commentary/ed090806b.cfm), Rea S. Hederman, Jr., September 8, 2006, The Heritage Foundation

[17] Kenworthy, L. (1999). Do social-welfare policies reduce poverty? A cross-national assessment. *Social Forces, 77*(3), 1119-1139.

[18] Bradley, D., Huber, E., Moller, S., Nielson, F. & Stephens, J. D. (2003). Determinants of relative poverty in advanced capitalist democracies. *American Sociological Review, 68*(3), 22-51.

Further reading

- Shepard, Jon; Robert W. Greene (2003). *Sociology and You* (http://www.glencoe.com/catalog/index.php/program?c=1675&s=21309&p=4213&parent=4526). Ohio: Glencoe McGraw-Hill. pp. A-22. ISBN 0078285763.
- Alan Gillie, "The Origin of the Poverty Line", *Economic History Review,* XLIX/4 (1996), 726

External links

- History of the U.S. Poverty Line (http://eesc.orst.edu/agcomwebfile/edmat/html/em/em8743/part1/officialgovt.html) by Tom Gentle, Oregon State University.
- United States Department of Health and Human Services Poverty Guidelines, Research, and Measurement (http://aspe.hhs.gov/poverty/index.shtml)
- 2007 United States Department of Health and Human Services Poverty Guidelines (http://aspe.hhs.gov/poverty/07poverty.shtml)
- Debraj Ray 1998, *Development Economics*, Princeton University Press, ISBN 0-691-01706-9.
- World Summit for Social Development Agreements (http://www.un.org/esa/socdev/wssd/agreements/index.html), United Nations
- The "elimination" of poverty (http://www.inclusivedemocracy.org/journal/vol4/vol4_no1_takis_poverty.htm), Takis Fotopoulos, The International Journal of Inclusive Democracy (http://www.inclusivedemocracy.org/journal/), vol.4, no.1, (January 2008).
- Fisher, Gordon (2005-12-16). "Relative or Absolute — New Light on the Behavior of Poverty Lines Over Time" (http://aspe.hhs.gov/POVERTY/papers/relabs.htm). Department of Health and Human Services. Retrieved 2008 -01-16.

Article Sources and Contributors

Channel Lake, Illinois *Source*: http://en.wikipedia.org/w/index.php?title=Channel_Lake%2C_Illinois *Contributors*: ChannelLake, Droll, Dual Freq, Funandtrvl, Nyttend, Pearle, Plastikspork, Ram-Man, Seth Ilys, TheCatalyst31, 3 anonymous edits

Census-designated place *Source*: http://en.wikipedia.org/w/index.php?title=Census-designated_place *Contributors*: AbruCH, Ahoerstemeier, Ajd, Alansohn, Amikake3, Arjayay, Ary29, Averette, Betacommand, Bkell, Bkonrad, CRGreathouse, Chasdenonno, CryptoDerk, Cybercobra, Dale Arnett, Dreadstar, Eaglizard, Elkman, EncMstr, Everything counts, Freddiem, Friedo, GeorgeLouis, Gioto, Gparker, H-stt, Iampreston, Jacooks, Jiang, John Vandenberg, Jojocool117, Joseph Solis in Australia, Jpers36, Kb3edk, Kbdank71, KeithH, Lajsikonik, Leslie Mateus, Lexusuns, Lights, LouI, Luxem, Lwalt, MacTire02, Mako098765, Mateo SA, Michael Hardy, Mightymights, Moreau36, Motthoangwehuong, Musiphil, Niklem, Nyttend, Odin of Trondheim, Olivier, Orlady, OverMyHead, Pax:Vobiscum, Pmanderson, Polaron, Poop in the toilet!, Qrc2006, Reddi, Ricky81682, Rsduhamel, Schicksal, Sdw25, So-nj, Stevertigo, Supertouch, The Thing That Should Not Be, Thewodakpa, Tillwe, Uncia, Velvet elvis81, Vmenkov, Vrysxy, WhisperToMe, Wik, Wikip rhyre, Yellowdesk, Zzyzx11, 37 anonymous edits

Antioch Township, Lake County, Illinois *Source*: http://en.wikipedia.org/w/index.php?title=Antioch_Township%2C_Lake_County%2C_Illinois *Contributors*: Kranar drogin, Omnedon, Rich Farmbrough, TheCuriousGnome, 2 anonymous edits

Lake County, Illinois *Source*: http://en.wikipedia.org/w/index.php?title=Lake_County%2C_Illinois *Contributors*: AshyLarry, Astuishin, Barek, Barry1776, BrianRBurgess, Brohmn, Burpelson AFB, CanisRufus, Cg41386, Cherry1000, Chris24, Cliff1911, Coyau, D6, DS1953, Danielba894, Danny, Dingar, Docu, Doprendek, Dralwik, Dual Freq, Fishal, Funnyhat, Geomaven, Gongshow, GrahamHardy, Hephaestos, Hoejoehoe, Hu12, Javierpuertorico, Johnpacklambert, JustAGal, Kimyou, Kranar drogin, Libertyville, Lightmouse, Mac'ero, Mjvan12, Monegasque, Mrmaniac3, Mtq777, N4yeli, Nyttend, Omnedon, Onyourside, Peter200, Ram-Man, Rich Farmbrough, Rjhanson54, Rrostrom, Saros136, Sasha Chhh, SchuminWeb, Shsilver, Slightlyslack, Smallbones, Sortior, Speciate, Template namespace initialisation script, Terryn3, The Interior, The Person Who Is Strange, TheCatalyst31, TheCuriousGnome, Tim!, Ufwuct, Ugur Basak, Voyager, WhisperToMe, Worldenc, Yahel Guhan, 99 anonymous edits

United_States *Source*: http://en.wikipedia.org/w/index.php?title=United_States *Contributors*: "Country" Bushrod Washington, (, (aeropagitica), *hollywood*, -Barry-, -Majestic-, 0, 041744, 067pansaol, 08OceanBeach SD, 08murtaghkc, 09LFIRTH, 0Bammer, 0ts0, 1.21 jigwatts, 1028, 10nn, 119, 123Mike456Winston789, 1251thestrokes, 12835S, 1297, 149AFK, 1658, 16@r, 172, 18alex12, 19994best, 19starwars77, 1oddbins1, 1of3, 2005-03-04T23:39Z, 2005-03-09T21:41Z, 2005-04-08T13:18Z, 212etc, 23prootie, 28421u2232nfenfcenc, 2nd Piston Honda, 2t56, 3 years too many, 3232330, 400, 50imp, 5ptcalvinist, 621PWC, 66.81.31.xxx, 6r6r6muyr6uh, 7i5, 8r1ck, A Clown in the Dark, A Softer Answer, A State Of Trance, A Train, A bit iffy, A s williams, A-b-c-d-e-f, A-giau, A520, AAA765, ACD605, ACSE, ACW, ACupOfCoffee, AED, AEMoreira042281, AIRBUS, AKGhetto, AKMask, ALEXF971, AP1787, ARC Gritt, ARCoolguy, ARUenergy, ASDFGH, ASOTMKX, AStudent, AVM, AZCactus1, Aaker, Aalahazrat, Aang-kai, Aaporter 87, Aaron DT, Aaron Einstein, Aaron Schulz, Aaron mcd, Aaron7chicago, AaronS, AaronY, Abayarde, Abc is a number, Abce2, Abdalla A, AbeOlsen, Abecedare, Abilityfun, Abonazzi, Abtract, Abuchman1, Academic Challenger, Acalamari, Acanz, Accurizer, Ace ETP, Acerperi, Acetic Acid, Acio83, Acjelen, Acmilan15, Acroterion, Adam Bishop, Adam Krellenstein, Adam Pamilih, Adam1213, AdamFirst, AdamJacobMuller, AdamRetchless, Adamski24, Adanedhel21, Adashiel, Adavidb, Addicted04, Addisonn., Addit, Adelina and Hannah, Adhall, AdjustShift, Adko110, Adolphus79, Adoniscik, Adpete, Adraeus, Adriaan, Adrian, Adrimaster3, AdultSwim, Aenar, Aeonx, Aesopos, Aeyl13, Afinebalance, Afropuffs71, Agdaman4life, Agtaz, Ahlabonde, Ahmad.ibn.as.Sayyid, Ahmadyk23, Ahodes1, Ahoerstemeier, Ahuskay, Aidepolcycne, Aido2002, Aim Here, Aiman abmajid, Aimjeff2, Airdish, Aislar, Aivazovsky, AjaxSmack, Ajwitte, Akamad, Akanemoto, AkvinRox, Akyoyo94, Al Villarruel, Al-Andalus, Al1encas1no, Alakasam, Alan Flynn, Alan.ca, AlanBarrett, Alandavidson, Alanhaggai, Alansohn, Alarics, Albert0057, AlbertR, Albinostorm, Alcuin, Aldie, AlefZet, Alegoo92, Alephcat, Alex S, Alex43223, AlexJP, AlexNebraska, AlexOUK, Alexander Vince, Alexdragon, Alexignatiou, Alexihlo, Alexnovo, Alexsautographs, Alexwcovington, Alfiboy, Alhutch, Ali Karbassi, Ali a, Ali944rana, Aliazimi, Alientraveller, Alienus, Alilhappything, Aliwaseem, Alkivar, All Male Action, All in, AlltimeJourneyFan, AllyUnion, AlmasterGM, Alodyne, AlphaEta, Alphabetagamnma, Altanar, Altenmann, Alvin Lee, Alvinrune, Amberrock, Amcfreely, AmericaIsNumberOne, AmericanXplorer13, Americanization, AmiDaniel, Amikeco, Aminatam, Amirb1, Amp873, Amren, An Siarach, AnOddName, Anakus, Anaraug, Anastrophe, Anativecantonesespeaker, Ancheta Wis, Andersmusician, Andkore, Andonic, Andre Engels, Andrea moro, Andrejj, AndresTM, Andrevan, Andrew Levine, Andrew Steller, Andrew c, Andrew0921, Andrewbarber79, Andrewlp1991, Andreworkney, Andrewphelps, Andrewpmk, Andrewsthistle, Andris, Android79, Andy Marchbanks, Andy-ch, Andy120290, AndyBQ, AndyZ, Andyabides, Andyb215, Andychen86, Andypandy.UK, Andypham3000, Andyso, Anetheron589, Angel David, Angela, Angelastic, Angelboy, Angelo De La Paz, Angr, AngryParsley, Angryhobo16, Angusmclellan, Anhydrobiosis, Anilocra, Animum, Aninnymouse, Anirvan, Ankhet, Ankit jn, Ankitsingh83, Ann O'nyme, Anna Lincoln, Annama29, Annandale, Annasalia, Anon6666, AnonMoos, AnonyLog, Anonymity001, Anonymous editor, AnselmoLacerda, Anshumsaxena, Antandrus, Antarticstargate, Anthere, Anthony, Anthony Appleyard, Anthony717, Anti-Wuss, AntiBodies, AntiBones, AntiVanMan, Anto475, Anton27, AntonioB, Antopi, Anubis2051, Anujill, Anywikiuser, Anzoooo, Aogouguo, Ap, Apatens, ApolloBoy, Apostrophe, Appleboy, Applemaster, ApprenticeFan, Aptdwn26, Apwvt, AquaD, Arad, Araiderplaya, Aranel, Aranherunar, Aravind V R, Arbuge, Arcadie, Arch dude, Archaelicos, Archaic, Archanamiya, Archimedez, ArchonMeld, Arcimpulse, AreJay, AreaControl, ArglebargleIV, AriGold, Aricci526, Aris Katsaris, Aristophanes68, Aristotle1990, Arkalochori, Armenia81, Armyrifle9, Arnon Chaffin, Arnoutf, Arpingstone, Arsenio, Arshadm, Art LaPella, Artifactblue, Artimbo, Arwack, Arwel Parry, Arx Fortis, Aryavib, Asbestos, AscendedAnathema, Asderoff, Asdfjkl123, Asenai, AshAloy, Ashenai, Ashershow1, Ashnard, Ashrichards90, Aspensti, AspiringScholar, Assyria 90, Asterix70, Astolfo Appleyard, Astonvilla91, AstroNomer, Astronautics, Astrotrain, Atchom, Atemperman, Athenean, AtilimGunesBaydin, Atlant, Atmans, Atomique, Atoric, AuburnPilot, Auburnirishman, Audacity, Aude, Aufheben, Auroranorth, Ausimus, Austen Redman, Austen rules, Austin Hair, AustinKnight, Auswiger, Author782, Autonova, Avala, AvalonXQ, Avargasm, Avenue, Avg, Avicennasis, Avidius, Avsrule1921, Awilcox1533, AxSkov, Axel Svensson, Axeman89, Aximill, Aydchery, Ayla, Ayrton Prost, AzaToth, Azmurath, B. Fairbairn, B0at, B0mb3d, B1link82, B64, BD2412, BECASC, BG7, BGManofID, BGordon, BJCenters, BMF81, BRG, BSveen, Baa, Babablacksheephaveyouanywool, Babbage, Babbby, Babygrand1, Babylicious2, Backslash Forwardslash, Backwalker, Badanagram, Badbe0, Bahn1k, Balderdash707, Ballboys10, Ballestrain, Bambuway, Bamford, Banders234, Bandity, Banes, Bansp, BanyanTree, Baplexed, Barbara Shack, Barbieblader, Barek, Barfooz, BaronLarf, Barrylb, Barryob, Bart133, BarzaniKurd16, Basbalfrk, Baseball Bugs, Baseballbaker23, Bash RF, Basilisk4u, Basohead, Bassbasketball03,Assplr19, Bastique, Battem, Baxter 10, Bayonetblaha, Bazonka, Bbatsell, Bbenjoe, Bbert1994, Bbert1994cchs, Bbx, Bchalfin, Bdesham, Beach bum7, Beamathan, Beamer103, Beano, Bearingafish, Bedford, Beef cakes, Beefy247, Beepoppab, Beetle1042000@yahoo.com, Beezer137, Before My Ken, Beh-nam, Behemoth98, Beland, Belligero, Ben Arnold, Ben D., Ben Lunsford, Ben76266, BenFrantzDale, BenH, Benbristol, Benc, Benhocking, BenjaminTsai, Benjiboi, Benkalt, Benlisquare, Benson85, Bentobias, Benwbrum, Berek Halfhand, Bet1973, Beta m, Betmeno, Bevo, Bevo74, Bhadani, BhaiSaab, Bhamv, Bhe22, Bibbledebobblede, Bibbly Bob, Bidgee, Big Adamsky, BigBoyRubio, BigFishy, Bigairboarder121, Bigbear bh, Bigbluefish, Bigferdes, BiggKwell, Bigmak, Bigmanly, Bigredbrain, Bihco, Bilby, Bill McCloskey, Bill william compton, Bill-on-the-Hill, BillC, Billheller, Billim1, BillyBoom, Billymac00, Binkhaldun, Biohazard930, Bird1455, Birdmessenger, Birdofprey, Birthofthehive, Biruitorul, Bishonen, Bisteca, BitWarrior, Bjarki S, Bjmarfito, Bjpremore, Bkell, Bkkeim2000, Bkonrad, Bky1701, Black Kite, Black and White, Black lyzard, BlackHades, Blackanthrope, Blackcell16, Blackmongoose, Blackshafen, Blainster, BlankVerse, Blarg2.0, Blastwizard, Blaylockjam10, Bleff, Blessthishouse, Blindman shady, BlisteringFreakachu, Blog Mav Rick, Blondace, Blu Aardvark, Blue7545, BlueAg09, BlueSalo, Bluedog423, Bluedogtn, Bluegold, Bluemask, Bluemoose, Bluwiki123, BlytheG, Bmdavll, Bmgoau, Bmicomp, Bmitchelf, Bo, Bob A, Bob rulz, Bob schwartz, Bob swatsky, Bob121, Bobblehead, Bobblewik, Bobbob, Bobbum 47, Bobby Mouth of the South, BobbyAFC, BobbyJ, Bobbyi, Bobchalk, Bobdobbs1723, Bobdylan3589, Bobfreshwater, Bobisbob, Bobmack89x, Bobo The Ninja, Bobo192, Bobobobo, Bobozz, Bobtheuglyhobo, Bogdan, Bohemian tosspot, Boman1990, Bondegezou, Bonewah, Boneyard90, Bongwarrior, Bonked116, Bono983, BonsaiViking, Boogaryou, Boomshadow, Boothy443, Boraczek, Bornintheguz, Bosonic dressing, BostonRedSox, Bowei Huang, Bowei Huang 2, Boxbrown, BozMo, Bporter28, BradM, Bradifer, Bradwart, Braffit, Brainboy109, Brainhell, Branddobbe, Brandmeister (old), Brandon, Brant4, Brat32, Bratsche, Bravada, Bread for the ducks, Breisblonde22, Brendan O'Shea, Brendanconway, BrendelSignature, Brendenhull, Brennan626, Brenton, Brian Pearson, Brian the Editor, Brian0918, Brianga, Brickfield, Brighterorange, Brilliburger, Brion VIBBER, Brisvegas, Britannic1, Britcom, British legend, BritishWatcher, BritneySpearsAddict, Britsrule06, Brockert, Brodey, Broken Segue, BrokenSegue, Brokenfuse, Brossow, BrotherFlounder, Bruguiea, Brutaldeluxe, Brutannica, Bryan Derksen, BryanG, Bryant693, Bssc81, Btipling, Buaidh, Buddha24, Buffalob07, Bugs5382, Bull Market, Bulldog123, Bullshark44, Bully25, Bunny-chan, Burningview, Burntsauce, Burroughsks88, Burto88, Bustoff, Butros, Buzz-tardis, Bwinn, Bwithh, C. M. Harris, C.Fred, C.Löser, CALR, CAMorris, CASportsFan, CDHgrün, CDN99, CJ, CJ Aikman, CJ King, CJK, CJLL Wright, CKBrown1000, CLW, CMSimons, COF, COMPFUNK2, CORNELIUSSEON, CPacker, CQJ, CRouleau, Cacophony, Cactus.man, Cafemusique, Cafzal, CalJW, Calcprgmer10011, CalicoCatLover, Califunya, Calliopejen1, Callmemrshowoff, Calltech, Calmypal, Calvin Lourdes He, Cam24, Cam275, Cambrasa, CambridgeBayWeather, Camenzind, Cameron, Cameron12345, Cameron12355, Camr, Camw, Can man138, Can't sleep, clown will eat me, Canada Jack, Canadaisawesome, Canadia, CanadianCaesar, CanadianLinuxUser, Canderson7, Caniago, CanisRufus, Canofmeat, Canterbury Tail, Cantus, CapitalR, CapitalSasha, Capmo, CaptainNate, Captaincoach92, Carbonite, Cardiff123098, Cardinalwraith, CaribDigita, Carlaude, CarlosMarti123, Carnogaunt, CarolinianJeff, Carowinds, Carrp, Carthradge, Casper Gutman, Castletower, Castncoot, Catdude, Catswilltakeovertheworld1day, Causa sui, Cause of death, Caveman 07, Cbrett, Cbrown285, Cburnett, Ccacsmss, Cchow2, Ccs4ever, Cdc, Cdcon, Ce garcon, Cea801, Cecil722, Cecilgol, Cedric88, Cedrus-Libani, Celestianpower, Cello0wnsy0u, Cellurl, Celyndel, Censusdata, CentralTiger74, Centrx, Cenwin88lee, Cerealkiller13, Cerebrus13, CesarB, Ceyockey, Cfkane84, Cfquinlan, Cgoodell, Ch'marr, ChKa, Chadlupkes, Chanlyn, Channelone, Chanting Fox, ChaosSorcerer91, Chaosdruid, Chaosname, Charlie chaplin, CharlotteWebb, Charruss, Chason11, Chatori, Che829, Cheatcode2, Check two you, CheckPlus, Chedorlaomer, Cheese bombs, Cheese007, Chessman, ChesterTheWorm, Cheung1303, Cheungpat, Chiamus, Chic-ism, Chickenofbristol, Chicopac, Chillum, Chindia (China-India), Chinju, Chinneeb, Chipmunkdavis, Chivista, Chobbs138, Chochem, Chochopk, Chocoforfriends, Cholmes75, Chongkian, Chris 73, Chris Mason, Chris cufc, Chris-dotcom, Chris432, Chrisahn, Chriscoooley, Chrislk02, Chrismichener1, Chrisranders, Christian Historybuff, Christian List, Christinam, Christinewales, Christopher Kraus, Christopher Mulcahy, Christopher Parham, Christopher higham, Christophore, Chrose, Chunky Rice, ChunkySoup, Chzz, CieloEstrellado, Cienade, Cieran 91, Ciroa, Cirt, Citizen-of-wiki, Cityhighkid09, Civil Engineer III, Cje, Cjeiler, Ckatz, Clab6, Claidheamohmor, ClamDip, Clanget, Classic Fan, Clawson, Clayoquot, Cleared as filed, Clicketyclack, Clickpop, CliffC, Cloakdeath, Clorox, Closedmouth, CloudNine, Cloudreaver, Cma, Cmadler, Cmc0, Cmcginnis, Cmguy777, Coaster1983, Coasterlover1994, Cocoaguy, Cocoliras, Cocoma, Cod1337, Code Logic, Codingmonkey, Codu, Cofax48, Cohesion, Cold Light, ColdFusion650, Collier, Colonel Cow, Comayagua99, Cometstyles, Commander Keane, Commonbrick, Comput65, ComputerBuker, Computerwhiz171, Comrade Martin, Concordia, Coneslayer, Confiteordeo, Congolese fufu, Conk 9, Conman23456, Connelly, Conningcris, Connormah, Conor 33, Conquistador48, Contemptiblepunx, Contra10, Conversion script, Conveyance, Conza1994, Cool Blue, Cool Stuff Is Cool, Cool-guy357, Cool4x411, Coolalec11, Coolcaesar, Cooldoug111, Coolhawks88, Coolio Grooverider, Cooliocheck1, Cooljuno411, Coolpeng123, Coolstorybrooo, Copperchair, Coraje, Cordless Larry, Cordova, Coredesat, Corgimaster, Corlen, Cormac Canales, Cornellrockey, Corriebertus, Corticopia, Corvun, Corwin8, Cosmic Latte, CosmologyProfessor, Costlab, Cougar Draven, Coulraphobic123, CoulterTM, Countersubject, Countmall, Courcelles,

CovenantD, Cow, Cowman109, Cpavlovski, Cprompt, Cptbuck, Cr0ssfire, Crad0010, Craigy144, Craky, Crash14, Crazy8jacky, Crazycomputers, Crazyparadigm, Crazytail2, Crazytales, Crd721, Credema, Creez34, Crevaner, Criptofcorbin, CristianoMacaluso, Critical.solvent, Crm18, CrnaGora, Crobzub, Crohall, Cronic 5085, Crowbaaa, CrucifiedChrist, Crunchedfor6, Crusadeonilliteracy, Cruzian, Crxssi, Cryptic, CrypticBacon, CryptoDerk, Cryptosquid, Crzrussian, Cs-wolves, Csigabi, Csurguine, Cthia, Ctjf83, Ctryellowbird, Ctuavocado, Ctyankeeboy1638, Cuak's, CuboneKing, Cuchullain, Cuimalo, CumbiaDude, Cunado19, Curps, CurtisJohnson, Cutiewithabooty123, Cutrupi04, Cuye, Cvcdrk, Cverlo, Cyan, Cyanidethistles, Cybercavalier, Cybercobra, Cyde, Cynical, Cynicism addict, Cyp, Cypher056E, Cyril Washbrook, Cyrilthemonkey, Cyrus111, D prime, D-Rock, D.M. from Ukraine, D.illah, D6, DARTH SIDIOUS 2, DBaba, DC, DCB4W, DCGeist, DEW 7, DGJM, DHN, DJ Clayworth, DJHeartAttack, DL79OL, DLJessup, DMCer, DMellis, DTSX, DVD R W, DaGizza, Dabbler, Dablaze, Dabomb87, Dabrown, Dacoslett, Dacres, Dadude3320, Daedalus969, Dah31, Dak is back, Dale Arnett, Dale-DCX, Dalek, Dalstadt, Damian Yerrick, Damicatz, Dammit, Dampe98, Dan Tarrau, Dan100, Dan56, DanGalligan, DanKeshet, DanRF, DanShearer, DangerousBob, Dangoodman, Daniel, Daniel J. Leivick, Daniel Morales, Daniel Quinlan, Daniel4sw, DanielCD, Danieldis47, Danielfarzan, Danieljo2013, Danielsavoiu, Danigoni, Dankie, Danny 17, Danoasis, Dans tes rêves, Dante2308, DanteAgusta, Dantesqueman, Danthemankhan, Dar-Ape, Daredude, Dargen, Dark Shikari, Dark glaive, Dark12334312, DarkAudit, Darknshadow, Darkwind, Darrenemerath23, Darth Panda, Darthbob100, Dasani, Dave Runger, Dave.e.bondeson, DaveSDCali, Daveharvey83, Daven200520, Daveswagon, David H Braun (1964), David Johnson, David Kernow, David Levy, David s graff, David.Monniaux, DavidFarmbrough, DavidLevinson, Davidcannon, Daviddamoore, Davideus86, Davidgrazi7, Davidstrauss, Davidyoung, Davila, DavyJones14, Db099221, Dbabbitt, Dbachmann, Dbenbenn, Dbiehl, Dbtfz, Dc freethinker, Dcammz, Dcandeto, Dcflyer, Dday14044, Ddlamb, Ddssaa, Ddye, De Moriarty, Deacon of Pndapetzim, Deanb, Deanos, Deathmelon92, Deathperson, Deathregis, Debate789, Deborahhosen, Decumanus, Deeceevoice, Deep impact, Deepstratagem, Deeptrivia, Deetdeet, Defective 13, DefendEurope, DeifieD, Deli nk, Delirium, Delj523, Delldot, Delta x, DeltaSpectre, Deltabeignet, Demomoke, Demon-lad, Demonix, Den fjättrade ankan, Denelson83, Denihilonihil, Dennis44, DennisAro, Deor, Der4, DerHexer, DerMeister, Deranged bulbasaur, Derek Ross, Dereks1x, Descendall, Destructoman, Deus Ex, DevastatorIIC, Devil1337, Deville, DevinCook, Dewan357, Dewy9, Dfarmer, Dffgd, Dfrg.msc, Dgilman, Dharmabum420, Dhn610, Dhruvhemmady, Diberri, Dickrichards, Diderot's dreams, DieYuppieScum, Differentgravy, Digitalme, Dillan.Murray, Dillon29, Dilweg, Dina, DinajGao, Dinopup, Dipper2, Dirtjacket, Discordanian, Discospinster, Dismas, Dispenser, Dissident, Distal24, Diverse82, DividedByNegativeZero, Dizzyizzy, DjDark1 2009, Djinn112, Dkdantastic, Dlohcierekim, Dlwldud96, Dmartin969, Dmeranda, Dmerrill, Dnheff, Dniete97, DobbyD, Doc Quintana, Doc glasgow, DocKino, Docu, Doczilla, Dogma5, Dogs2000, Dolemite17, Dom188816, Domaleixo, Dominic, Dominics Fire, Domminico, Domthedude001, Donarreiskoffer, Donsiku, Donut2, DoomsDay349, Dooney 30, Doopokko, Doprendek, Dorfner, Doringo, Dostluk, DoubleBlue, Doug Funny, Doug Johnson, Dougdorval, Doughboy, Dougweller, Doverspa, Dp462090, Dpaajones, Dpark, Dpm12, Dposse, Dpotop, Dpr, Dr kapp 89, Dr.alf, DrVerlucci, Draconas, Draconix, DragonCurse, DragonFlySpirit, Dragonfir731, DragonflySixtyseven, Dragonsscout, Dragoon1025, Draicone, Drakky, Dralwik, Dran0259, Dranzini, Drcornwallis1, Drdyer, Dreaded Walrus, DreamGuy, Dreamafter, Dreish, Dres, Drestros power, Driftwoodzebulin, DriveMySol, Drmies, Drtjumper, Druid1090, Drumguy8800, Drummingboi, Dsaddsa, Dsfsdf, Dthdc4, Dtobias, Dubayuw, DubbaG, Ducker, Dude018219293, Dudewheresmypizz4, Dudtz, Dufekin, Dukakis1988, Duke Hazzard, Dumaka, Dunro, Durin, Durova, Durruti, DustFerret, Dustimagic, Dustintml, Dutchdrummer99, Dvdcdr, Dwilso, Dwrcan, DylanW, Dynam1te3, Dynamicknowledge28, Dysprosia, Dzulco, Dzzl, Długosz, E Pluribus Anthony, E Pluribus Anthony redux, E30e, EH74DK, ESkog, Eagle4000, Earl Andrew, Earpol, Earth, Eastlaw, Easyid, Ebakunin, Ebolito, Eclecticology, Economiesofscale, EconomistBR, Ecopirate, Ectoplasmer, Ed Poor, EdBever, Edcolins, Edgar181, Edgars2007, Edgeshappy, Edgeweb, Editor18, Editor19841, Editore99, Edmundbaird, Edobrichoo, Edsanville, Eduardo Sellan III, Edward, Edward Z. Yang, Edward nz, EdwinHJ, Edwy, Effer, Eggeggeggegg, Egil, Eileen's Dream, Eink21, Einzig27, Eivindgh, Ek8, Ekcrbe, Ekrub-ntyh, El C, El Generico, El Slameron, El kulu ngile, ElBenevolente, ElTyrant, Elbbom, Elduncarno, Electionworld, Electromud, Electrosoccertux, Elfguy, Eli81993, EliasAlucard, Elitistnerd, Ella Plantagenet, Elliskev, Ellomate, Elockid, ElockidAlternate, Eloquence, Elpato, Elpaw, Eluchil, Eluchil404, Emanuel Kingsley, Emcee2k, Emigrant85, Emilson, Emorph, Emperors Harbinger, Empion, Emsko, Emx, Encephalon, Encheart, EncycloPetey, Energizer07, Energyfreezer, Enkrates, Entirelybs, Enviroboy, Eoghan888, Eoin123, Epbr123, Epeefleche, Epicadam, Epiovesan, Equilibrium007, Erachima, Erfemf, Erginer, Eric outdoors, Ericamick, Ericci8996, Ericwirtanen, Erudy, Erwetr, Esimal, Eskimo Boots, Esrever, Essaregee, Eswiki, Etaoin, Etcher, Ethan c.00, Ettrig, Eu.stefan, Eugen Simion 14, EugeneZelenko, Eurocopter, Eurofire, Euroleague, Evans1982, Evb-wiki, Evensteven1, Evercat, Everyking, Evhs00, Evil Monkey, Evil saltine, Eviloldman, Evolauxia, Excimer3.141597, Existensorg, Exlibris, Extra999, Extremalnik, Eyas Hajeh, Eyu100, Ezhiki, Ezratrumpet, F. H. Black, F20ljc, FAH1223, FCStaehle, FCYTravis, FDR, FESTORM 5, FEastman, FF2010, FFLaguna, FOARP, FSF-Rapier, FT2, Fabartus, Fabhcún, Fabrictramp, Facts707, Fadeddreams5, Faelomx, Faezdel, Fagnostic, Failofbeaner, Faithlessthewonderboy, Falconclaw5000, Falcorian, Fame, Fangz, Fantrl, FargoWells, FarnhamJ, Fastily, Fatbug1, Fatla00, Fatpeople12345, Favonian, Fctk, FeanorStar7, Fearthesun, Featuresaltlakecity, Fedallah, Feinoha, FelisLeo, Fenice, Fennec, FeralitYzERO, Ferio252, Ferkelparade, Fernirm, Ferrariman, Feydey, Feyer, Feysaln, Fezzaga, Fibo1123581321, Fidelpmn2n, Fieldday-sunday, Fiet Nam, Fifty7, Filby, Filemon, Finoman40000, Firebug, Firehose, Firemancvn72, Fishal, Fishiehelper2, Fitzy2280, Fixman, Fkarim, Flackson, Flag-Waving American Patriot, Flamarande, Flamealchemist, Flashflash;, Flatterworld, Flaunted, Flexure, FloNight, Floaterfluss, Flockmeal, Flogiz, Flooey, Flosssock1, Fluent aphasia, Flurry, FlushinQwnzNyc, Fnfd, Folksong, Following specific instructions whispered by a mysterious cat, Fonduelazone, Fonzy, Football123213, Footballfan190, ForemanDomai, Former user 6, Forsyth.joel, Fortdj33, Fourthords, Fowlance, Fplay, FraDany, Fragsta, FrancisTyers, Francs2000, Frangibility, Frank, Frank0051, Frank94.kane, FrankCostanza, Frans Fowler, Frazzydee, Freak in the bunnysuit, Freakofnurture, Fred Bauder, Fredbauder, Fredericoxxx, Fredrik, Fredy.00, Freepenguin, Frencheigh, FreplySpang, FreshBulletTime, Freyr, Freyr35, Frichmon, Friedo, Frjlove, Frogsprog, Frostyservant, Frullic, Fshoutofdawater, Func, Fundamental metric tensor, Funnyhat, Funsides, Furdeen, Future2008, Fuzheado, Fuzzygenius, Fuzzyslob, Fvincent, Fvw, G Rose, G. Campbell, G.W., G2bambino, G8summit, GCarty, GD, GHe, GOP904, GRBerry, GSTQ, Gabbe, Gabrielsimon, Gadfium, Gaff, Gagi, Gailbawlz, Gaius Cornelius, GaiusTimiusAwesomus, Gaiusknight, Gakon5, Galati, Galoubet, Galwhaa, Gamaliel, Gamer007, GamerMod, Gamersedge, Ganesha1, Ganeshk, Ganrither, Garfieldt, Garfld986, Garkeith, GarnetRChaney, Garret Beaumain, Garretgwozdz13, Gary King, Garynine, Garytmlee, Gasolene, Gator1, Gatsby4life, Gauss, Gaz, Gazpacho, Gbd1995, Gdo01, GearedBull, Ged was here91, GeeZee, Geeoharee, Gegik, Gekaap, Gene Nygaard, GeneralDuke, Genghiskhanviet, Gengis, Genius Chimpanzee, Gentgeen, Geo Swan, Geo0910, Geo247, Geofat, Geoff Plourde, GeoffreyVS, Geoking66, GeometryGirl, George, George415, GeorgeFormby1, Georgeadgdgdgdwngo, Georgia guy, Geothermal, Gerald Farinas, Gerg1031, Germanlight, Gershwinrb, Getcrunk, Getoar TX, Gfnb, Gggh, Ggunit, Ghalas, Ghermit, Ghost1552, GhostPirate, Gianfranco, Giftlite, Gika, Gilgamesh, Gilhun, Gilliam, Gillydude, Gimferrer, Gimligimli1, Gimmetrow, GinaDana, Gingey933, Gintong Liwanag Ng Araw, GiovBag, Gjking, Gkmx, Gku, Glaucus, Glaxxo1, Glenn, Glenrqz, Globeism, Gman124, Gnaughton, Gnevin, Go229, Goalie1998, Gogo Dodo, Goheels36, GoingBatty, Gojoshg, Golbez, Gold Index, GoldRingChip, Goldblue21, Goldom, Gomarlins, Gomm, Good Olfactory, Good friend100, GoodDay, GoodSirJava, Goodandevil, Goodoldpolonius2, Goosehumper, Gop 24fan, Gopman1, Gordonfan, GorillazFanAdam, Goswamir14, Gr8opinionater, GraemeL, GraemeLeggett, GraemeMcRae, Graft, Graham87, GrahamTM, Gramaic, Grampion76, Grandin, Granpuff, Grant.Alpaugh, Graphic, Grapplequip, Graufbold, Green Giant, Green01, Greenman, Greg96786, GregAsche, Gregcaletta, Gregorio, Greier, Grendelkhan, Gretzkyv99, Grey Shadow, Grey Wanderer, Greyhood, Grick, Grig541, Grimhelm, GrittyLobo441, Grjfiridkeokiski8j, Grm wnr, Grod028, Grop, Grosplant, Grosscha, Ground Zero, Grstain, Gruguuru, Grundle2600, Grunny, Grunt, Gsarwa, Gshave, Gsingh2300, Gtuthtnfgurnfu, Guanaco, Gubbagubbaba, Guettarda, Gugilymugily, Guilherme Paula, Guitar 01, Gullit Torres, Gumbogumbo, Gurch, Gurchzilla, Gurlpleezer101, Gurubrahma, Gustavb, Guuzjo, Guy M, Guðsþegn, Gvhgu, Gwenham, Gwern, Gwernol, Gwybedyn, Gyre, H0ttamale2, H1478971, H5+R1A, HAHS 25, HIZKIAH, HJ32, HTBrooks, HU Pilot, Habsfannova, Hadal, Hadrian89, Hadseys, Haemo, Hager jeff, Hahachuckles, Haham hanuka, HaireDunya, Hairy Dude, Haisthater, Hajor, Halberdier25, Halcionne, HalfShadow, Halibutt, Hall Monitor, Hallows AG, Hamedog, Hammer1980, Hans Dunkelberg, Happy-melon, HappyVR, Happyme22, Hardyplants, Harej, Hargrimm, Harkell, HarrietTubman, Harriganesh, Harro5, Harryboyles, Haruo, Harvestdancer, Hateless, Hateusa11, Haukurth, Haverberg, Havoc 8790, Hawaiian717, Hayden43, Haydenstone2, Hcheney, Hchrishicks, Hcnorris, Hdsijhfoiuahfyusdaof, Hdt83, HeadCaptain, Headbanger756, Headlikeawhole, Heart Like a Wheel, HeartofaDog, Hearts of clear lake, Heat3000, Heaviestcat, Heby73, Hectorlf, Heegoop, HeikoEvermann, Heimdal, Heisencat, Heitor CJ, Hejsansvejsan, Hektossus, HeleneSylvie, Helios solaris, Helix84, Helixblue, Hellcat fighter, Hello14, Hello32020, Hello5959us, Helz0415, Henrik (old), Henry Delforn, Henry Flower, Henrygb, Hephaestos, Hequals2henry, Herdrick, HereToHelp, Hereforhomework2, Hernlund, Herp Derp, HerrGoebbles, Herunar, Herzie, Hexlor, Hfarmer, Hgspring, HiEverybodyHiDoctorNick, HiLo48, Highfields, Highvale, HillChris1234, Hipocrite, Hippiedude0, HippoOwl, HisForeShadow, Historian19, Historic Phoenix, Hiuby, Hk1992, HkCaGu, Hmains, Hmghosthost, Hobbentodd, Hobbesdude, Hobojoeandco, Hohum, Hoi jeon moo sool, Hojimachong, Hole62, Holledgeh, HollyAm, Holy Ganga, Homestarmy, Homeworld1031tx, HoodedMan, HooperBandP, Hoovernj, Hoping To Help, Hoponpop69, HoratioVitero, Horses In The Sky, Hoshie, Hosterweis, Hot Stop, Hotelcalifornia1976, Hotshots2006, Hottentot, House1090, Househelps, Housesitter, Howard Alexander, Howcheng, Howdybob, Hseldon10, Hu, Huadpe, Huangdi, Hudude, Huduoroi, Huelenrite, Hugo Chavéz, Hugo999, Human diversity, Humanproject, Huphelmeyer, Hurricane111, Hurricanefloyd, Husond, Hut 8.5, HuuhManFui, Huzzahmaster018, Hydrogen Iodide, Hypathia, Hyperscope, Hystlen Caeth, Héous, I-2-d2, IANVS, IJA, ILguy06, ILuvAmerica, IMOW, IPodAddict181, IRelayer, IZAK, Iakane49, IamBeastly20, IamYossarian, Iamthejake2000, Iamwisesun, Ian Dunster, Ian13, IanBailey, Iandiver, Ianhatch, Iankap99, Ibagli, Icairns, Iceberg3k, Icecradle, Iced Kola, Icedevil14, Id447, Idaltu, Idib, Ief, Ifrit 1780, Igilli, Ignacio Egea, Igo4U, Ihatepotsmokinghippies, Ikanreed, Ikcyzrteip, Ike9898, Ikh, Ikyork, Ilikepie2221, Illegitimate Barrister, Illuminatiscott, Iluvmesodou, Ilya1166, Ilyanep, Imaginary heroes, Imaninjapirate, Imbilly, Immunize, Impactplayer, ImperatorExercitus, Imperium Europeum, Improper Bostonian, In Defense of the Artist, InGearX, InShaneee, Inanthropomorphism, Inchead, Indexheavy, Indopug, IndyLawSteve, Infidelis, InformationalAnarchist, Infrogmation, Inimino, Inka 888, Insommia, Insuperable, Integralolrivative, Intelati, Intellibeing, Intelligentlove, Intelligentsium, Interfector, Internet quickie, Intimidated, Intmotinf, Introman, Intuitionz, Inwind, Ionutmortu, Iota, Iridescent, IrishHermit, Irishguy, Irishpyro, Irixman, Irrypride, Iruss14, Irviding11, Isawooty, Itai, ItalianGreyhound, Italianboy10, Italiano111, Italo, Iudaeus, Iuio, Ivankinsman, Ivannoriel, Ivejustbegun, Ivirivi00, Ixfd64, Izehar, J Clear, J Di, J intela, J. Daily, J. Finkelstein, J.delanoy, J3ff, J4bin, JACOB911, JAHL14, JB50000, JBH23, JDooley, JEN9841, JForget, JG2K6, JGHowes, JHMM13, JHP, JHarrelson, JHunterJ, JIP, JLMadrigal, JLaTondre, JMBZ-12, JMS Old Al, JPArnett, JPD, JPMcGrath, JStewart, JTBX, JTM, JVz, JW1805, JWSchmidt, JZZDD, JaGa, Jaberwocky6669, Jabolba, Jacda1313, Jacek Kendysz, Jacj, Jack.Hartford, JackDayton, JackLumber, Jackbirdsong, Jackdude101, Jacketpook, Jacksoncw, Jacksonsimon29, Jacky man Toronto, Jacob Lundberg, Jacob.clarke19617, Jacobah, Jacobolus, Jacottier, Jadabocho, Jafd88, Jak2, Jaked92, Jakro64, Jalada, Jalfredl, Jambalaya, Jamcib, James A. Stewart, James.d.mccollum, JamesMLane, JamesMadison, JamesMaxwell, JamesReyes, JamesTeterenko, Jamesday, Jamesmarkhetterley, Jamespeterka, Jamie C, Jamiemason, Jammy-dodger, Jamyskis, JanCeuleers, Jane tyra, Janechra, Janilson1995, Janviermichelle, Jaranda, Jared mckittrick, Jarheadigor, Jarjarbinks10, Jarkeld, JarlaxleArtemis, Jasgbro, Jason Potter, JasonAQuest, Jasoncward, Jasondmath, Jasonfitz, Jasper Deng, Jatkins, Java13690, Java7837, Jawsman2, Jaxad0127, Jaxbobby, Jaxl, Jay, Jayafrancis, Jaydee928, Jayhawk08, Jazz214, Jazz457, Jbloun1, Jc900, Jcm4, Jcmenal, Jcrocker, Jcw69, Jd027, Jdavidb, JdeJ, Jdforrester, Jdjoz0098, JeLuF, Jecar, Jecowa, Jedi master laybourn, Jeepday, Jef018, Jeff G., Jeff3000, JeffTL, Jeffgoin, Jeffrey S, Jeffz, Jengod, Jenks24, Jennavecia, Jennica, Jensboot, Jerahad, Jeremy221, JeremyA, Jeremygbyrne, Jeronimo, Jerry Jones, JerryFriedman, JerrySteal, Jerzy, Jess567, JesseRafe, JesseW, Jester2001, Jetlagorange, Jeyt5, Jez, Jezpuh, Jfarhou, Jfjkdc, Jfrancis, Jglees, Jgritz, Jguk, Jguk 2, Jh51681, Jhendin, Jhsounds, Jiang, Jibran1, Jihad!, Jim Douglas, Jim T, JimIrwin, JimWae, Jimbo D. Wales, Jimmy Kustes, Jimmy Slade, Jimmy butt, Jimmy-barnes, JimmySlade, Jimmyhead521, Jimothytrotter, JinJian, Jitse Niesen, Jivedevice, Jj137, Jjacobucci, Jjb, Jjhcap99, Jkbena612, Jkfp2004, Jklin, Jkschroeder27, Jleon, Jlujan69, Jm51, Jmabel, Jman1974, Jmbrock, Jni, JoanneB, Jobe6, Jobroluver98, Jodafire, Joe Jarvis, JoeBlogsDord, JoeSperrazza, Joebieber, Joeblakesley, Joehaer, Joel27, Joelr31, Joelv40, Joesmoe8, Joesr55, Joeyramoney, Johan91, Johann Wolfgang, JohannVII, John, John Fader, John Quincy Adding Machine, John Smith M.D., Ph.D, John Torn, John-1107, JohnDBuell, JohnOwens, JohnRuskin, JohnWittle, Johnian144, Johnleemk, Johnmc318, Johnmike81, Johnowenlangham, Johnstone, Johntex, Johntinker, Jojhutton, Joker123192, Jokeroler, Jon Doh, JonMoore, Jonarooni, Jonas Viper, JonasanRat7, Jonathan Drain, Jonathan-Morris711, JonathanFreed, Jonathunder, Jondel, JonesMama, Jongpil Yun, Jonnie d smith, Jonsafari, Jonyyeh, Jooler, Joolz, Joowwww, Jophus00, Jorcoga, Jordanian forever, Jordz, JorgePeixoto, Jorgezoltan, Jose77, Joseph Buchanan, Joseph Solis in Australia, Josh Parris, Joshdboz, Joshua Scott, Joshua368, Joshurtree, Jossi, Josteinaj, Jowe189, Joy, Joylifen, Joyous!, João Neves, Jpers36, Jpgordon, Jprulestheworld, Jrboi, Jrdioko, Jredmond, Jrkarp, Jrkso,

Jrn, Jrockley, Jrt989, Jrtayloriv, Js00902, Jsoltman, Jtdirl, Jtkiefer, Jtokarsbr, Jts6293, JuJube, Juantay, Judgement, Judgesurreal777, Judygreenberg, Julain Barbarosa, JulieADriver, Jumbuck, Jumple, Junes, Jupiter Green, Juppiter, Jusdafax, Jusjih, Jusme1, Just a member, Justanotherguitarist, Jwissick, Jwking, K. Lastochka, K157, K50 Dude, KBtoys09, KF, KFan II, KGBarnett, KGasso, KL22, KS-U92, KTC, Kaasje, Kablammo, Kadelia, Kafziel, Kaio-ken x10, Kaisershatner, Kajal4, Kajmal, Kaldari, Kalsermar, Kangy, Kansan, Karam.Anthony.K, Kardrak, Karimarie, Karl, Karl Dickman, Karukera, Kasyapa, Katalaveno, Katefan0, Kathmandu2007, Katydidit, Kaveh, Kazikame, Kazu-kun, Kazvorpal, Kbdank71, Kbolino, Kedar63, Kedi the tramp, Keebler71, Keegan, Keegster5, Keelm, Keilana, Keith H., Keithd, Kellen`, Kelly Martin, Kelstonian, Kelvinc, Kember, Kemet, Kemiv, Kenaldinho10, Kennard2, Kenndo, Kennethduncan, KensterFox, Kerrysfrench, Kertof, Kesla, Kesselring88, Kethinov, Kev2310, Kevin B12, Kevin Myers, Kevin Taylor, Kevinmon, Kevinp2, Kevjumba, Kevs, Kewp, Kg vasilev, Kgrr, Khalid, Khanada, Khatores, Khatru2, Kiddster93, Kier07, Killdevil, Killer Swath, Kimon, KimvdLinde, Kinaro, King Dracula, King Gemini, King of Hearts, King wiston, KingBob1, KingTT, Kinggeedorah, Kingpin13, Kingturtle, Kinneyboy90, KinturkMan, Kirby of Doom, Kirchherr, KirchnerE07, Kire 1123, Kirill Lokshin, Kirkieboy, Kishorekumar 62, Kitch, Kjaer, Kkm010, Klaam, Klafubra, Klander Brigade, Klaser, Klonimus, Klr390, Kmacsimus, Kman543210, Kman665, Kmanblue, Kmaster, Knife2sharp, Knightshield, Knorpel, KnowledgeOfSelf, Knowsitallnot, Knoxbury, Knutux, Koavf, Kochamanita, Kogsquinge, Konstantin, Korora, Kotjze, Kotniski, Koyaanis Qatsi, Kozuch, Kpjas, Kr1st1deejay97, Kraalg, Krakaet, KrakatoaKatie, Kralizec!, Krator, Krellis, Krich, Krismanx, Krizpykremekid, Krupo, Kschwerdt514, Ktsquare, Kttreasure, Kueldude, Kukini, Kulikovsky, Kumioko, Kunaygi, Kungfuadam, Kurieeto, Kurtle, Kuru, Kurykh, Kusma, Kuzco Mii, Kuzmaxster, Kwame1234, Kwamikagami, Kwijiborjt, Kwirky88, Kwizzard, Kyh87, Kyle Barbour, Kyle J Moore, Kyle sb, Kyle1278, Kyle4321, Kzzl, Kızıl Şaman, L33th4x0rguy, L3p3r, LGagnon, LIbertyInSpace, LOL, LaMenta3, Labby24, Labeledstickmen, Labz, Lachatdelarue, Lacrimosus, Ladb2000, LadnaV, Laguna72, Lakshmix, Lala0000, Lalein91, Lampoonsvacation, Lan56, Lanboney, Langston, Laomei, Lapsus Linguae, Lasindi, Last king of Frisia, LaszloWalrus, Latics, Latitude0116, Lauboo, Laubz83, Laucheehung, Laudaka, Lauren, Laurinavicius, Lawe, Lawyer2b, Laxbrain0110, Lazyandy123, Lcarscad, LeAviateur537, Lear 21, LeaveSleaves, Lec domicilarias, Lectonar, LedRush, Ledzeppelin321295, Lee Daniel Crocker, Leeannedy, LegitimateAndEvenCompelling, Lego872, Legoboy2000, Legokid, Lehi, Leif, Leithp, Lemonaderox, Lemonsawdust, Lenoxus, Lenzar, Leodj1992, Leon7, Leonard G., Leroyinc, Leszek Jańczuk, Leuko, Levineps, Lewisgs, Lexor, Lfgnyc, Lfh, Lhademmor, Liber, Liberalcynic, Liberi Fatali37, Liechten11, LifeStar, Liftarn, Lightdarkness, Lightmouse, Lightningfro, Ligulem, Lijealso, Likeminas, Lilac Soul, Lilpinoy 82, Lima bean of the north, Limideen, LincolnSt, Lindberg, Linuxbeak, Lipton1995, Lisa517, Lisazipper, LittleDan, LittleRoughRhinestone, Littlelance, Livajo, Living001, LizardWizard, Ljriccio9j, Llakais, Llec, Lloyd.smith, LlywelynII, Lmusielak, Lneal001, Locaracle, Lockesdonkey, Logan, Lokibolp, Lolnoraj, LonelyMarble, LongBay, Longhorns4life89, Longwalkshortpier, Loodog, Looper5920, Looxix, Lord Emsworth, Lord Geznikor, Lord Moostyer, Lord Terminus, Lord Voldemort, Lord Wikipedian of the Wiki, Lord alexandre, LordHarrison, LordSimonofShropshire, LordSnooze, Lordflamel, Lordvaluemart, LoreleiLynn, Lorenzop, Lorenzosabbadini, Loser11111111122222, Lotje, Lou1986, LouI, Louis Waweru, Lousyd, Louturks, LovesMacs, Lowellian, Lowreeder, Lowride, LoyolaDude, Lrusso99, LtWinters, LuGiADude, Luby, Lucas Duke, Ludvikus, Luiscolorado, Luisztdt, Luk, Luke334, Luke56, LukeM212, Lukemadness, Lukep913, Luna Santin, Lupin, Lupita°, Lupo, Lussier, Luxor99, Lycurgus, LynnMeilleur, Lynntoniolondon, M a s, M2MM4M, M5891, MAG Fry, MBisanz, MC MasterChef, MCRTGP, MDCore, MECU, MER-C, MJCdetroit, MJDTed, MJK, MKers06, MONGO, MPS, MWaller, MWatson15, MZMcBride, Mabbass, Mac Davis, MacGyverMagic, Macfan93, Macg4cubeboy, Macintosh User, Macjaeh, Mackay 86, Mackensen, Mackoy20, Macphisto12, Madbiggie00, Madcoverboy, Maddiekate, Madman445, Madskillz225, Maelnuneb, Maestro25, Magicman710, Magister Mathematicae, Magister Scienta, Magnet paper, Magog the Ogre, Maheshvenk, Maijinsan, Mailer diablo, Mailman9, Mais oui!, Majoreditor, Majorly, Makaristos, MakeChooChooGoNow, Malarious, Malcolm, Malhonen, Malo, Maltmomma, Mana Excalibur, Mandey7, Manicsurge, Manix, Manop, MantisEars, Manutdglory, MarSch, Marcfromns, Marcika, Marco Neves, Marco polo, MarcoTolo, Marcoescobar, Marcopolo7890, MarcusMaximus, Marcusliou5, Marcworca, Mareino, Marielleh, Marine(uc), Mark Antony666, Mark Renier, Mark Ryan, Mark Sheridan, Mark83, MarkDalit, MarkDeyarmond, MarkSweep, Markpravda, Marmote, Marnanel, MarsInSVG, MarsRover, Marshall Williams2, Marshmallow00, Marshman, Marskell, Martarius, Marth2k, Martin S Taylor, Martin451, MartinHarper, Martinense, Martinp23, Martinwilke1980, Martious, Martyman, Martynrss2, Marvolo Gaunt, Marymer, Marysunshine, Mashboy, Mason.Jones, Massimo Catarinella, Master Jay, Master and Commander, Master of Puppets, Master of the Oríchalcos, Mastercomputerpro9999, Masterjamie, Masterpiece2000, Matchgame12345, Mateo SA, MathewDill, Mathgod333, Mathpianist93, Matt 8, Matt Crypto, Matt Gies, Matt Hales RDFC@hotmail.com, Matt Yeager, MattPDG, Mattabesta, Matthewprc, Mattkidd12, Matt296, Mattworld, Matty 1988, Mattyh190, Mattyhar, MaulYoda, Mav, Max rspct, Max(is not)well, Max20020012002, Maxatl, Maximus Rex, Maximuspower, Maxkoyama, Maxkramer, Maxmc, Maxpower37, Maxsht9, MaxxFordham, Mb93j, McDogm, Mcanmoocanu, Mcattell, Mceder, Mcimo88, Mckyj57, Mcmoor, Mcsweet, MeThOd, Meco, Medvedev40, Meegar, Meelar, Meeptastic, Meeve, MegX, Megaandroid30, Megahmad, Megan 189, Megapixie, Megawattbulbman, Mehicdino, Melsaran, Member, Menchi, Mendel, Merbabu, Merlot70, Mermaid from the Baltic Sea, Merovingian, Mesgul82, Messiisking, Metafisyx, Metehan1992, Mets501, Metz2000, Meursault2004, Mevlüt Kılıç, Mexihcatlacatl, Mghabmw, MghtyRoach, Mgiganteus1, Mh10190, Mhesselb, Mhking, MiKellie, Michael Essmeyer, Michael Glass, Michael Hardy, Michael Patrick, Michael-Billa, Michael.Urban, Michael2-, MichaelBillington, Michaelpkk, Michaelwuzthere, Mickey gfss2007, Mickeykozzi1, Mickeymickey11, Mightymights, Mihas, Mika293, Mikael Häggström, Mike Christie, Mike Halterman, Mike Pehipol, Mike Rosoft, Mike nofx, MikeBravo, MikeLynch, Mikefzhu, Mikehawk, Miken32, Mikeroodeus, Mikestern2, Mikeybadass05, Mikhailov Kusserow, Mikinator, Mikpass1, MilFlyboy, Mileage, Milk of magnesia, Milk the cows, Milkmandan, Millahnna, Million Moments, Millsdavid, MindscapesGraphicDesign, Minesweeper, Minesweeper.007, MiniZiper, Minjamonkeyw00t, Minna Sora no Shita, MinnesotanConfederacy, Miranda, Mirrorkirby9062, Mirv, MisfitToys, Miskin, Miss tylie, Miss-jessie-gal, Missionary, Missmarple, Missouri Rebel, Mister Alcohol, Misterrick, Mistertheman, Misza13, Mitaphane, Mitchumch, Mitsukai, Mitth'raw'nuruodo, Mixcoatl, Miyokan, Mizuphd, Mjf3719, Mkweise, Mlazik, Mm40, Mmarch31, Mmavipc, Mmx1, Mmxx, Mnlira013, Mnmazur, Mo0, Moby-Dick3000, Moderate2008, Modest Genius, Modulatum, Moe Epsilon, Mogden, Momentary, Momo san, Momus, Mon amour est avec la France, Moncrief, Mongoletsi, Monkeyblue, Monkeyman, Monkeywrenchjim, Mono, Monobi, Monsieur Fou, Montrealais, Montyofarabia, Moondog88, Moonraker0022, Mordien, Moreau36, Morhange, Morphh, Morton devonshire, Morwen, Mosca, Mosemamenti, MosheA, Mostgiganticlyingmouthofalltime, Mosthauntedjp, Mostly Rainy, Motthoangwehuong, Mouloud47, Moustan, Moverton, Moxy, Mozartmuzic, Mprudhom, Mpscan, Mr Adequate, Mr Stephen, Mr Taz, Mr Twix, Mr. D Vega, Mr. bwn, Mr.Do!, MrD9, MrDrew508, MrGRA, MrJanitor1, MrMick, MrRadioGuy, Mrbelvedereposter, Mrbrown, Mrjpb104, Mrknowitman, Mrld, Mrmaroon25, Mrtobacco, Mrwojo, Mrzaius, Mschiffler, Msludwick, MtnBrook, Mufka, Mulad, Mumble45, Munulu999, Muraad kahn, Murali nmv, Murchy, Murdocc, MurphysLawn, Mushroom, Musicalantonio, Musiphil, MuskokaON, Musse-kloge, Mustanglover, Muya, Mwanner, Mx3, Mxcatania, Mxn, Myanw, Mydreamistofly, Mygeneration01, Myleslong, Myriad, Mystaker1, Mystify85JEC, Mário, Mäximus Rex, N-edits, N-true, N00bschoolbus, N244888, N5iln, N734LQ, NBS525, NByz, NGerda, NHRHS2010, NJOrlando, NJguy281, NKT45, NSLE, NSR, NW036, NWOG, NYKevin, Nach0king, Nachomania, Nadavspi, Naddy, Nader85021, Nakon, Nappyrootslistener, Narco, Naryathegreat, Nascar1996, Nat, Nat Krause, Nat682, Nate1012, NateXMS, Natgoo, Nathan8225, Nathanalex, Nathanbolt, NativeForeigner, Nattuvijay76, Natty4bumpo, NatusRoma, Naus, Nautilator, NawlinWiki, Nbatra, Nblc, Nbrzycki91, Nd24, Nduhn101, Neal Finne, Nebulousity, Nehrams2020, Neilbeach, Nelandquinten, Nelfer, Nelro, Nelsondavidd, Neo233, NeoLamagra, Neoballmon, NeonNight, Neondiscogod, Neorunner, Neovu79, Nerdalert101, NerdyScienceDude, Nerothanes, Netbenefit, Netesq, Netkinetic, Netoholic, Netscott, Neuroscientist, Neutrality, Nevada, Nevilley, New England, NewEnglandYankee, Newstormer, Nezbie, Nhakimi, Niceley, Nichalp, Nichofas, Nicholas F, Nicholas Tan, Nick, Nick123, Nick4gwen, NickMartin, Nickptar, Nickyaberdeen, Nico, Nicolas Barbier, Nicolewasherexx, Nicolharper, Nidam, Nige, Nigeria01, Nighend, Nigholith, NightDragon, Nightscream, Nihonjoe, Nika, Nikodemos, Nikofeelan, Nikola Smolenski, Nikoniko, Nikostar, Nikwad244, Nil Einne, Nil105994, Nimbley6, Nine999999999s, Nipplemaster5000, Nippoo, Nirvana888, Nishishei, Nissi Kim, Niteowlneils, Nitharsank, Nitya Dharma, Nivix, Njoedits, Nkrosse, Nmpenguin, No Guru, No toco at aol dot com, Noleander, Nomad, Nomist, Nonenmac, Noozgroop, NorCal764, NoraBG, Northmeister, Northwales4u, Northwest, NortyNort, Norwikian, Nosimplehiway, Nosskyline, Not a dog, NotAbel, Nothlit, NovaNardis, Novidmarana, Novis-M, Nowhither, Nrafter530, Nrana99, Nroscha, Nscheffey, Nskrill, NuKkEm, NubKnacker, Nuberger13, NuclearVacuum, NuclearWarfare, Nufy8, Nuge, Nukebros, Numberoneconti, Numberonesurvivor, Numerouno13, Nunh-huh, Nvw, Nxxos, Nydas, Nzzl, OBrienMadHouse, ONEder Boy, OOODDD, Obaidz96, Obhave, Obi-WanKenobi-2005, Obli, Oblivious, Octane, Oda Mari, Offenbach, Offrecord09, OhitstragikX, Ohmygodeven, Ohnoitsjamie, Okiefromokla, OlEnglish, Olaf Eriksson, Old Guard, Old Moonraker, Old Right, OldRight, OldRightist, Oldag07, OldakQuill, Oldmankdude, Ole10589, OleMaster, Oleg Alexandrov, Olegwiki, Olivier, Ollieegg, Olorin28, Omarlegend, Omarmorek, Omega9380, Omicronpersei8, Ommnomnomgulp, Omnibus, Once in a Blue Moon, One, One man orgy, OneDoubleO, OneEuropeanHeart, OnePt618, Onetwo1, Onixz100, Onopearls, Onorem, Oo64eva, Oobopshark, Opaldragon1, Opelio, Openthedoor754, Operation Spooner, Optim, OptimumPx, Optional Toaster, OrbitOne, Ordealbyfire, OreL.D, OregonD00d, Oreo Priest, Orthorhombic, Oscarthecat, Oskart, Ost316, Otashiro, Otebig, Otolemur crassicaudatus, OuroborosCobra, Outoforderuk, Ovcp1, Owen, OwenBlacker, OwenX, Oxboomer, Oxford Comma, Ozi, P.W.Lutherson, P0per, PAWiki, PCock, PDH, PFHLai, PL290, PRRfan, PRSteve27, PS2pcGAMER, PZFUN, Pabejar, Paco8191, Pah246, Pahistorian, PainMan, Paine Ellsworth, Paix Pan, Pak21, Pakaran, Paleo dragon, Paleoblues, Pamri, Panfakes, Panserbjorn, Pantherskin, Papercrab, Para, Parachnidox, Paradise7, Parallel or Together?, Paraside, Parliamentcity, Parmadil, Parthasarathy B, ParvatiBai, Pastorsson93, Patapsco913, Patioco, Patka, Patman2648, Patrick, Patrickneil, PatrikR, Patsson77, Pau07, Paul A, Paul August, Paul Tyrrell, Paul.h, Paul2.0, PaulHanson, Paulinho28, Pauljeffersonks, Paulmcguane, Paulsprecker, Pax:Vobiscum, Pb30, Pbedford, Pcman22594, Pcpcpc, Pcstico, Peak, Pearle, Pecher, Pedant, Pedantic Turkey, Peeweeteehee, Pegasus1138, Pekin Republican, Peloneous, Pengkeu, Penny1020, Penser, Penshurstpride, Penubag, Pepper, Pepsi132, Pepsidrinka, Per Honor et Gloria, Peregrine Fisher, Perfect Editor, Perfect77, PerfectStorm, Perfectionaintperfect, Perkinator 3000, Permethius, Perseus, Son of Zeus, Persian Poet Gal, Peruvianllama, PeteJayhawk, Peter, Peter L, Peter.C, PeterK, Peterale4444, Peterlin, Pethr, Petrb, Petrux, Pewwer42, Pfortuny, Pgan002, Pgecaj, Pgk, Ph.D.Nikki, PhJ, PhaedoHD, Phantomsteve, Pharos, Phe, Phil Boswell, Phil5329, Philanthropy, Philip Baird Shearer, Philip Stevens, Philip Trueman, Philip200291, Phillies1fan777, Philliesfan035, Phillip Fung, Phillip J, Phinnaeus, Phoebe, Phoenix2, Phoenix7777, Photo me2004, Phrankle, Phrasia, Phxcesar, Pi399, PiMaster3, Picapica, Pidgeontoes, Pierco, PierreAbbat, Pigman, Pilko182, Pill, Pillow2011, Pilotguy, PinchasC, Pinotgris, Pioelad, Piss420, Pizzadeliveryboy, Pkpat2011, Plainsong, Plapre09, Plastic, Plasticbadge, Plusgood, Pm06420, Pmackarate, Pman31014, Pmanderson, Pnatt, PoccilScript, Poetaris, Poilil, PokeYourHeadOff, Polaron, Polisci, Politepunk, Politis, Polylepsis, PonileExpress, Poojean, Poopymatt, Poor Yorick, Poppa Bear, Porkman385, Porsche997SBS, Portillo, Postco, Postdlf, PotentialDanger, Pperos, Prasenberg, Pratheepps, Premeditated Chaos, Preschooler.at.heart, Preslethe, PrestonH, Pretty Green, Preuninger, Primalchaos, Primarypenguin, PrinceForte, Prodego, Producercunningham, Prospect1, Proteus, Pryde 01, Pschemp, Pseudo-Richard, PseudoSudo, Psy guy, Pubber, Publicus, Puertorico1, Pulmonaryarchery75, Pupitar13, Purodha, Pushbreak, Pwilk, Pwjb, Pwnage8, PxT, Pyroclastic, Pyroechidna1, Pyukman, Pé de Chinelo, Q Ramona, Qasaqsuyu, Qmwne235, Qqqqqq, Qrc2006, Quadel, Quadell, Quadrochosis, Quaker24, Quako5, Quanticles, Quasipalm, Quazywabbit, Quberoot, Query, Questchest, Quijebo, Quill, Quinnculver, Quite vivid blur, Quixoto, Quizimodo, Quizimodo (usurped), Qviri, Qwe, Qwfp, Qxz, R'son-W, R-41, R-Joe, R0xorz, R9tgokunks, RA0808, RG2, RIPSAW1986, RJII, RJN, RMHED, RPIRED, RPlunk, RPlunk2853, RS085, RTG, RVD, RVJ, RWhite, RZ heretic, RaCha'ar, Rab777hp, Rachaelho50, Radarsnobles, RadicalBender, Radicalsubversiv2, RadioBroadcast, RadioKirk, Raegis, RafaelG, Rageinacage, Ragib, Rahlgd, Rahuldharmani, Railer 893, RainR, Rainpat, Raja Hussain, RajaNeela1993, Ral315, Ralph3000, Rama, Rama's Arrow, Ramanglass544, Ramazan!, Ran, RanEagle, RanchoRosco, RandalSchwartz, Rande M Sefowt, Random contributor, Randomblue, Randombob12345, Randomengine, RandorXeus, Randy Johnston, Randy Kryn, Rangermike, Rangerwave, Rangoon11, Ranveig, Raoul-Duke, Raprchju, Rarelibra, Ratemonth, RattleMan, Raul654, Ravedave, Ravenhull, RaviC, Ravikiran r, Ravlo2004, Raybob95, Rayc, Raymond arritt, Rayoflight278, Rb2911, Rbankson, Rcharding, Rctbone, Rd232, Rdrgz93, Rdsmith4, RebaFan1996, Rebecca, Rebellion101, Reboot, Reck345, Red rawker, Red65, RedChord, RedWordSmith, Redd Baron1, Reddi, Redeem, Redgecko5, Redkloud, Rednaxela, Redwolf24, Reeces904, Reedy, Reelcheeper, Refdoc, Regancy42, Regionalsimp, Reid A., Reid1867, Reillyg, Reimelt, Reinthal, RekishiEJ, Reliancepowercoin, Rembo74, Remiel, Renegadeviking, Renorei, Rentastrawberry, Republicanbetter, Republikaner, Reticulum, Retired username, Retireduser1111, Rev.Ision, Revmachine21, Revolución, Rex Germanus, RexImperium, RexNL, Rextrim, Reywas92, Rfc1394, Rgrigal, Rhaegar, Rhatsa26X, Rhobite, Rholton, Ric36, Rich Farmbrough, Richard Allen, Richard Blatant, Richard David Ramsey, Richard0612, RichardDavila, RichardF, Richie123098, Rick Block, RickK, Rickipedia127, Rickney, Rickvaughn, Ricky@36, RickyTheBest, Rickyrab, Riconoen, Rifecu, Rillian, Rimbaud, Riphamilton, RiseAbove, Risker, Ritchy,

Ritimakovej, Rittmeister, Rjairam, Rjd0060, Rjd2 Deadringer, Rje, Rjensen, Rjm656s, Rjwilmsi, Rkt2312, Rlm0710, Rlundi86, Rmerlos, Rmhermen, Rmhs15, Rmky87, Rob 06-07, Rob Shepard, RobTranter, Robbiemt, Robbo chappers, Robbro7, Robchurch, Robdurbar, Robert K S, Robert.Mcleod, Robert25, RobertG, RobertHuaXia, Robertkeller, Robertson-Glasgow, Robertvan1, Robfergusonjr, Robin63, Robster1983, Rocafella, Rocastelo, Rocawear1994, Rochabm, Rockentui, Rockstone35, Rockyobody, RoddyYoung, Rodrigogomespaixao, Rofl, Roger Arguile, Roger Gianni, RogerK, Rogered, Rogertudor, Rogueagent123, Rohawn, Roke, Roleplayer, Rollercoasterlord, Rollo44, Rolo75, Romanm, Rompe, RonH, Ronaldmolina20, Ronthewizard1994, Roosevelt Churchill, Rory096, Rory2259, Ros24s, Rose Garden, Rossenglish, Rosshterry, Rossnixon, Rotten, RoundDaWorld, RoverTheBendInSussex, Rox Da Box, Roxi2, Roxorzursoxorz, Roy Brumback, Roy da Vinci, RoyBoy, Royboycrashfan, Rreagan007, Rrhain, Rrie, Rrius, Rror, Rrouse, Rsm99833, Rstrug, Rto and buddy, Rtyq2, Rubber ducky222, Ruchi tspl, Rudjek, Rudolph.cm, Ruhrjung, Rumkles, Run26.2, Ruy Lopez, Rwolfe85, RxS, RyGuy17, RyanGerbil10, RyanParis, Ryanjunk, Ryanluck, Ryanmcdaniel, Ryland510, Ryulong, Ryz05, RickK, S.U.H.E.L., S2006causa, S4diStiC, S71elements, S91by, S@bre, SCEhardt, SDJ, SDY, SEWilco, SFrank85, SG, SJMNY, SMC, SNIyer12, SPUI, SS Division Wiking, SSZ, STARWARSGEEK, SWBJACMK, SWalkerTTU, SYSS Mouse, Saashaw, Saathoff37, Saberswordsmen1, Sabrown100, Sabs1285, Sadads, Sadasd, Sade, Saforrest, Sagaciousuk, Sahrin, Saigon76nyc, Saksham, Sakutdjica, Salsb, Sam Coutu-Oughton, Sam Korn, Sam1993, Sam3199, SamEV, SamForestell, Samantha555, SameOldSameOld, Samir, Sammo, Sampo Torgo, Samroundface, Samuel Blanning, Samuel Sol, Samuelsen, Samwb123, Sanantonioguy, Sandersonjoe229, Sango123, Sangre viento, Sannse, Sansvoix, Sarah, Sardanaphalus, SarekOfVulcan, Saric, Saros136, Sasquatch, Sauerjos, Sauve.d, Savefrance, Savidan, Sayden, Saypotatoes8, ScAvenger, Scammerboy227, Scanlan, Scarab12, Scarian, Sceptre, Schadenfreude52, SchfiftyThree, Schmloof, Schnitzelman, Schoolsurfer, Schrödinger's cat, SchuminWeb, SchutteGod, Schutz, Schzmo, ScientificDude, Scientizzle, Scientus, Scientz, Scifiintel, Scifiradioguy, Scipius, Scirocco6, Sciurinæ, Scohoust, Scoobycentric, Scorchy, Scorpi0n, Scotchy500, Scott Burley, Scott Illini, ScottMainwaring, Scottmsg, ScriptBlue, Scrumy21, Sdgjake, Sdhonda, Sdnegel, Seablade, Sean, Sean Whitton, Sean gorter, SeanO, Seaphoto, Search4Lancer, Searchme, Seb az86556, Sebbe, Secret (renamed), Secret Saturdays, Secretlondon, Seeraamaazu, Sekolov, SemDem, Semiquaver, Sennsa, Senorquigles, Separa, Sepeople, Seqsea, Seresin, Seric2, Serlin, Serminigo, Serpentinite, SerpentsTail, Sesel, Sesu Prime, Seth Ilys, Settl746, Sevillano1, Sexxie grrl, Sexy sally, Señorplayero, Sfacets, Sfahey, Sfmontyo, Sgguitarist, Sgodth, Shade11sayshello, ShadowGuardian, Shadowboy667, Shadowjams, Shadowy Crafter, Shaggy2Dope, ShakataGaNai, Shaken Earth, Shalom93, Shame On You, Shanedidona, Shanel, Shaneomac8605, Shanes, Shanesterman, Shared Thought, Sharkticon Guy, ShaunES, Shawn m r, Shech736, SheikYerBooty, Shelleycakes, Sherzo, Shii, Shim jb, Shimgray, Shimonnyman, Shindleria235, ShipperS7, Shizhao, Shmew, Shoaler, Shoeofdeath, Shohei Okada, Sholtar, Shootappliances, Shoreranger, Shorne, ShortShadow, Shot007, Shpilo, Shreshth91, Shroudan, Shunt110, Shushruth, Sicamous, Sicilianshotgun, Sietse Snel, Sig0, SigPig, Sigma 7, Sigmaparadox, Signsolid, Silence, SilentOpen, Silver Sonic Shadow, Silverchemist, Simishag, SimonP, Simulcra, Sinmiedoanada, Sir Nicholas de Mimsy-Porpington, Sir Richardson, Sir smorgasbord, Sir48, SirGrant, Sirkevinalot, Siroxo, SivaKumar, Sj, Sjakkalle, Sjö, Sk.Sajjad, Skaakt, Skapur, Skarebo, Skate4me, Skateboarderf2, Skeeter08865, Skentelbery, SkepticBanner, SkerHawx, Skier Dude, Skizzik, Skomorokh, Skooing, Skullfiend, Sky Divine, Skyking360, Skylinegtrgsl, Skyoon, Skyring, Sladuuch, Slakr, Slashcoconut, Slatersteven, Slconnor uk, Slcut, Sleptrip, Sligocki, Slivertransplant, Sloman, Slowking Man, Slrubenstein, Slugmaster, Smaug123, Smcmanus, Smilemean, Smiloid, Smith03, Smith2006, Smoke., SmokeyVillain, Smokeythecat, Smr16, Smug Irony, Snailwalker, Snoozer347, SnowFire, Snowmanradio, Snowolf, Snoyes, Snpp, SoMuchTime, Soakologist, Soccerisawesome, Sodzei, Solipsist, Solitude, Somatosis, Some jerk on the Internet, Somedifferentstuff, Someone else, Somercet, Somethingoranother, Sonicfire3000, Sonjaaa, Sonyray, Soooclose, Soosed, Sopher99, Sophie2895, Sopranosmob781, Soulpatch, Soulstice, Soumyasch, Soundslikealotofhooplah, South Bay, SouthernComfort, Southsydney, Soxwon, Sp33dyphil, SpNeo, Space25689, SpaceFlight89, Spaceboy492, Spangineer, Spark98, Sparrowman980, Spartan078, Spartanloser, Spebudmak, Speedboy Salesman, Spellmaster, Spencer195, Spettro9, SpiderMMB, Spiesr, Spiff, Spike105, Spikethehawk, Spinoff, Spiralhighway, Spitfire19, Spitzl, Splash, Splat, Spliffy, Sportsjam3, Spudtater, Spundun, SpuriousQ, Squash Racket, Squideshi, Squoups, Sri 1988, Srikantkedia01, Srishti1984, Srs, Srujanjoshi, Srushe, Ss2343, Ssilvers, Ssolbergj, Sswonk, St00pidmonkey72, Stabby Joe, Stallions2010, Stan Shebs, StanZegel, Staplegunther, StarScout0304, Stars999, Stasven, Statalyzer, StateCollege101, Static11, Statsagogo, Staxringold, Steel, Stemonitis, Stephen Gilbert, Stephenb, Stepshep, Stevage, Steve the Beach, Stevec240, Steven J. Anderson, Steven407653, Stevertigo, Stevietheman, Stevo1000, Stickee, Storasyster, Stormie, Stormscape, Strangepalefighter, Stratman07, Streryan, Strideranne, StringRay, Stroganoff, Strongbad1982, Strovolos01, Stubblyhead, Student7, Stuman12, Stupidstupidstupid, Stwalkerster, Subseven26, Substancep, Subsurd, Sudeepneelakantan, SudoGhost, Suffusion of Yellow, Sufi dog, Suisui, Sumahoy, Sumanch, Sunil060902, Sunquanliangxiuhao, Sunray, Sunshineofyourlove12, Supaman89, SuperDude115, SuperSmashBros.Brawl777, Supercoop, Supergeo, Supergodzilla2090, Superm401, Supersexyspacemonkey, Supertask, SupremeRuler, Surcouf, Surgemg, Suruena, Suryoye, Susurrus, Suto, Sverdrup, Swainstonation, Swamilive, Swarm, Swatjester, Swedish fusilier, SwedishConqueror, SwedishZeta, Sweetback, Swellman, Swid, Swimm212, Swissbrazilian, SwnyMac, Swollib, Swordhunter0690, Syndicate, Synthe, Syrthiss, Sysrpl, Systemshocked, Syvanen, Szajci, Szakharin, Szyslak, T-rex, T1980, T800m101, TEG24601, TEMcGee, TFNorman, TFOWR, THEemu, THEunique, TJ Spyke, TJDay, TJive, TMC, TOMMASO, TOO, TORR, TR5000, TSO1D, TShilo12, TTBMA, TUF-KAT, TV1981HS, TXLonghorn, Ta bu shi da yu, TaRiQ 8d8, Tabletop, Taborgate, Tachyon01, Taco325i, Tad.k.22, Tafinucane, Tai112, Taifarious1, Taikodrumfun, TakuyaMurata, Tangotango, Tanishh, Tannin, Tapir Terrific, Tapir2001, Tapirs65, Tardis, Targeman, Tarheelz123, Tariqabjotu, Tarret, Tasc, TastyCakes, Tavatar, Tawker, Tayyabbaig, Tazchook, Tbhotch, Tbone55, Tbonnie, Tbsea, Tburgtbone06, Tdadamemd, TeaDrinker, Team6and7, Techdawg667, Techguy118, Technocratic, Teddey, Tedickey, Tehmadhatter, Tekto9, TeleComNasSprVen, Teletoonmaniac, Telex, Telsa, Telstar2626436, Tempodivalse, TenOfAllTrades, Tenunda, Terebigemuwan, Terence, Terfili, Terlynn4, Terra Novus, Terrasidius, Teryx, Testes rodriguz, TexasAndroid, Texaswebscout, Texinian, Texink, Texture, Tezero, Tf157, Tfine80, Tgannon, Tha Nad 2007, Thadkobylarz, Tharnton345, That-Vela-Fella, Thatdog, Thatotherperson, Thatpacokid, The Catholic Knight, The Cunctator, The Disco King, The Epopt, The Evil IP address, The Evil Spartan, The Great Liberator, The Halo, The Hitch22, The Land, The Lizard Wizard, The Man in Question, The Minister of War, The Number Man, The Ogre, The Placebo Effect, The Punk, The Revert States of America, The Tetrast, The Tom, The Transhumanist, The Universe Is Cool, The Utahraptor, The Wingman Joker, The ed17, The last Wikibender, The lost library, The machine512, The monkeyhate, The slave to reality, The ultimator, The wub, The-Holy-Hand-Grenade-Of-Antioch, The1edit, TheArmadillo, TheDoctor10, TheFarix, TheFuzzyFive, TheGWO, TheGrappler, TheHoosierState89, TheKMan, TheMadBaron, TheNewPhobia, TheObtuseAngleOfDoom, ThePenciler, TheRanger, TheRomanLegion, TheWama, Theboondocksaint, Thebrother, Thechinese8, Thecrazydonvingigante, Theda, Thedec, Thedragonking, Thegargantua, Thegryseone, Thejason, Thelacker, Themane2, Themanwithoutapast, Themanwiththefish, Thematt523, Thenderson1, Thepebble3, Theredhouse7, Therequiembellishere, Theresa knott, Theriusthechronicler, Thermionic1, Thernlund, Therock40756, Thesevenseas, Thethirdperson, Theurgist, Thewolfstar, TheyMadeMeDoIt, Thezirk, Thingg, ThinkBlue, ThinkEnemies, Thisbites, Thiseye, Thomas Paine1776, Thomas S. Major, ThomasK, Thomasjamesfoster96, Thomasmallen, Thomasmatoushek, Threepwood89, Thricecube, Thryduulf, Thunder8, Thunderbird2, Thunderboltz, Thurifer, Tide rolls, TigerShark, Tigey, Tightguy400, TigranTheGreat, Tim Starling, Tim Thomason, TimBentley, TimShell, TimVickers, TimothyHorrigan, Timrem, Timrollpickering, Timwi, Tintero, TintininLisbon, Titch Tucker, Titoxd, Tiwonk, Tiësto, Tjl1128, Tjmayerinsf, Tjoe, Tjpark1111, TkerTimeSeeker, Tktktk, Tktru, Tkynerd, Tlogmer, Tma 88, Tmazhindu, Tobby72, Tobycat, Tobyk777, Tocino, Tom, Tom harrison, Tom-, Tom8866, TomStar81, Tomass288, Tombyt55, Tomdatom, Tomgibbens, Tominator93, Tomlillis, Tommybtennis, Tommybtennis2, Tommyjb, Tomwood0, Tomwsulcer, Tony Sidaway, Tony1, TonySt, Tonye104292, Tonywiki, Toonlad9, TopAce, TopazSun, Topbanana, Toropop, Tos, Tosfan4ever, Totla, Toussaint, Tparkes, Tpbradbury, Tpetross, Traal, Trainspotter, Trannyboi, Trasman, Trav230, Travelbird, Traversc, Travis Cleveland, TreXiaNos, Tree Biting Conspiracy, Tregoweth, Treisijs, Trevdna, Trevor MacInnis, TreyHarris, Tribulation725, Trinexx56, Trip Johnson, Tripacer99, Trontonian, Trottsky, Trovatore, Truecoder777, Truta1, Truthy not fakie, Ttwaring, Tubular, Tuckerdaman, Tud123, Turel, Turian, Turkuun, Tushar858, Tuspm, Tutmosis, Tuupu, Tverbeek, Twin Bird, Twirk88, Twump, Tx17777, Tycoon24, Tykell, Tyler McHenry, Tysalpha, U S Crusher, UBeR, UH Collegian, UberBicep, Ucanlookitup, Ucucha, Uebergeek, Ufwuct, Ugen64, Uglymoog, Uikos, Ulises Heureaux, Ultimus, Ultron, Umedard, Una Smith, Uncle J, Uncle Milty, Uncle Scrooge, Universe=atom, Unixxx, Unsc, Unsungzeros, Unyoyega, Uogl, Upper Lines, Uppie, UpstateNYer, Uris, Ursutraide, Uruguayo, Usaf2222, User F203, User2004, Usergreatpower, Usfirstgov, Utcursch, UtherSRG, Uuu87, Uv411, Uxejn, V6g3h7, Vacilando, Valentino, VampWillow, Van helsing, VanBurenen, Vandaltroll, Vaniba12, Vanished 6551232, Vanished User 3388458, Vanished895703, Vaquero100, Vargob, Vary, Vcb, VeLi33, Vectoor, Veeraane, Veertlte, Vega84, VegaDark, Vegaswikian, Veghead, Velho, Velvetron, Venkat ramkumar, Ventagro, Vera Cruz, Verdadero, Verrai, Verwoerd, VeryVerily, Verybigfish86, Viajero, Vicarious, Vicki Rosenzweig, Vikedal, Vilallonga, Vilerage, Villasmilraven, Vin Kaleu, Vincent Ricci, VinnyCee, Viogfernos, Violask81976, Violetriga, Virgile1991, Virgule82, Viridae, VitaleBaby, Vitoghf, Vitr0x, VividTara, Vladdraculdragon, VolatileChemical, Voltagedrop, Votekyle, Voyagerfan5761, Vrac, Vranak, Vrinan, Vsmith, Vulpesinculta51, Vuo, Vzbs34, W00kie, W123, W2ch00, WGee, WHEELER, WJBscribe, WLRoss, Wackojacko1138, Wackymacs, WadeSimMiser, Wafulz, Waggers, Wahuskies96, Waja, Walkiped, Waltersimons, Wambeter, Wangi, Wangry, Wangyunfeng, WarEagleTH, Warset, WartyX, Wassamatta, Wassupwestcoast, Wastful, Watch37264, WatsonE, Watti Renew, Wawat, Waxxup, Wayward, Wbush89, Weaponx 2.0, Weavale, Webdinger, WegianWarrior, Weien, Weinerheadshaboom, Wen Hsing, Wenteng, Werdan7, Weregerbil, Werideatdusk33, Wes!, WesleyDodds, Wester, Westnest, Wetodid, Weyes, Wfaulk, Whalleyj, Wharrel, Whatthree16, Wheathampstead Brko, Wheelboss, Where, Whiskey in the Jar, Whisky drinker, WhisperToMe, Whiteboywonder, Whitedwarf1.3, Whkoh, Who, Whodoesntlovemonkeys, Wickb55, Wickethewok, WideArc, Widefox, Wiggles (usurped), Wiglaf, Wiki alf, Wiki evan, Wiki man55, Wiki1609, WikiDao, WikiFlier, WikiLaurent, WikiWitch, Wikiacc, Wikibofh, Wikidemon, Wikih8lulz, Wikihistorian, Wikiklrsc, Wikipedian06, WikipedianMarlith, Wikipedianinthehouse, Wikipelli, Wikitigger, Wikiwert, Wikiwikifast, Wikizach, Wildinc, WildlifeAnalysis, Will Beback, Will Colwell, WillMak050389, WillOakland, WilliamThweatt, Williamscold, Willingham, Willtron, Wils4581, Wilson44691, WilyD, Wimt, Winkelthorp, WinterSpw, Wipe, Wisco, Wisconsinhockey, Wisdom89, Wise, Witchzilla, With goodness in mind, Withouthavingseen, Wjfox2005, Wkcp, Wkdewey, Wknight94, Wkstar, Wog274643, Wolf ODonnell, Wolfman, Wonderboy2, WoodElf, WoodenTaco, Woodstone, Woohookitty, Woozle, Work permit, Working for Him, World, World-os.com, Worldtimezone, Worrydream, Wrad, Wtmitchell, Wv235, Ww2censor, Wxlfsr, Wōdenhelm, XGustaX, XJamRastafire, XLerate, XU-engineer, Xaosflux, XavierGreen, Xblahx, Xen0blue, Xenus, Xezbeth, Xiggelee, Xkcd, Xkcer Man, Xlegiofalco, Xloveangelmusicbabyx, Xoder, Xolom, Xooxgmoneyxoox, Xrblsnggt, Xrobx921, Xtreambar, Xwoodandwater, XxTrillvillexX9, Xyzzyva, Y2kcrazyjoker4, YAKNOWDOUGAL!, YORKABE, Yahoowill54321, Yahyawahbeh, Yakuzai, Yama Thi Khuu, Yamamoto Ichiro, Yamla, Yamummasnotbad, Yandman, Yangon, Yaninass2, YankeeDoodle14, Yankees4919, Yanks724, Yanksox, Yansa, Yardalhirji, Yashgaroth, Yasis, Yath, Yeehyunc, Yelgrun, YellowMonkey, Yelyos, Yitzhak1995, Ylai, Yoganate79, Yohann4, Yojizu, Yongbyong38, Youhavebeentricked, Youndbuckerz, YourLawyer, Youssefsan, Ytmnd6, YuMaNuMa, Yubersoz, Yuckfoo, Yuje, Yupik, Yuri Gouveia Ribeiro, Yuusha, Z.E.R.O., ZICO, ZStoler, Zabararmon, Zach.vega1, ZachPG, Zachary8222, Zachlipton, Zachorious, Zafiroblue05, Zaksouthward, Zanimum, Zaphraud, Zarxos, Zauggb, Zayani, Zazaban, Zealotgi, Zeamays, Zebe, Zeldaluver100, Zenohockey, Zeppelin4life, Zeppelinrules, Zh, Zia demion, Zibzibzib, Zigger, Zloyvolsheb, Znarfaggle, Znode, Znorkel, Zoe, Zoicon5, Zoidy-Poo, Zondor, Zoney, Zotel, Zouavman Le Zouave, Zpowers, Zscout370, Zsinj, Zukariaa, Zvar, Zzuuzz, Zzyzx11, Zzzzzzus, Ævar Arnfjörð Bjarmason, Éponyme, सो ऽहम्, , , 5719 anonymous edits

United States Census Bureau *Source*: http://en.wikipedia.org/w/index.php?title=United_States_Census_Bureau *Contributors*: 007bond, Abductive, Acsenray, Ahoerstemeier, Al Lemos, Alan_d, Alex1011, Alexandria, AndrusTandrus, AnonMoos, Antandrus, Armonica, Astrocommander, Asturianu, Atlien, Aude, AutoGeek, Babij, Bender235, Big Adamsky, Bikepunk, Biruitorul, Bkonrad, Blargh29, Bonder00700, BradVanDam, Brandon5485, Buster7, CMG, CambridgeBayWeather, Can't sleep, clown will eat me, Cancandance, CapitalR, Chamal N, Chemboss, Chrism, Ckatz, Clayoquot, Clindberg, Closeapple, Cocircular, Coffee, Colonist, Conversion script, Csyberblue, Curps, Cybercobra, Darkfight, Daven200520, David Kernow, Davilaa, Davodd, Den fjättrade ankan, Deodar, Der Träumer (DE), Discospinster, Dmgaldi, Dough4872, Download, Dthomsen8, Duffman, EOBeav, ESkog, Eastlaw, Econr6, Eeaee, Epbr123, Epolk, EurekaLott, Evercat, Farquaadhnchmn, Favonian, Formulax, Framed0000, Frank101, Frazzydee, Furado, GABaker, Gadickson, Gidonb, Giftlite, Globalsolidarity, Gnominonwiki, GoldRingChip, Gpollock, Grafen, GregAsche, Ground Zero, Gugganij, Harro5, Hephaestos, ImperfectlyInformed, Izzy007 1, JONJONAUG, Jengod, Jerryseinfeld, Jholman, Joao Xavier, Joseph Solis in Australia, Jusdafax, Jvcdude, Kbdank71, KenFehling, Klaam, Koavf, Krellis, LOL, Lear's Fool, Leif, Leinad, Leszek Jańczuk, Levineps, LizardJr8, Lotje, Luna Santin, Marshman, Martinwj53,

Materialscientist, Mav, Meaghan, Michael Hardy, Microtony, Minesweeper, Musichman, Mxn, MyTwoCensus, Mysid, NickCT, NicoNet, Nightscream, NoSeptember, O18, Ohnoitsjamie, Olivier, One, Parkwells, Patrick, Pax:Vobiscum, Peak, Pearle, Pi, Pi zero, Pittimann DE, Pittimannl, Postdlf, Reach Out to the Truth, ReadQT, Rettetast, Revth, Ricav, Rich Farmbrough, RingtailedFox, Rklear, Robert Merkel, Rougher07, Rwwww, Sardanaphalus, Sarranduin, Scriberius, Semolo75, Seth Goldin, Shannon bohle, SimonP, Sk8tercolt18, So-nj, Spangineer, Srikeit, Srushe, Ssbohio, Stelmakia, Storm Rider, TJMcDJ, TOttenville8, Tanweer Morshed, Tedickey, Tenthfirst, The Epopt, The Nut, Themfromspace, ThinkingTwice, Tide rolls, Tim!, Tisane, Tom, Touchmytralala, TygerDawn, UH Collegian, Vanka5, Ventur, Wars, Wayne Slam, West.andrew.g, WhisperToMe, Willking1979, Yngvadottir, Yone, ZeWrestler, Zzuuzz, ش.ررز, 213 anonymous edits

Population density *Source*: http://en.wikipedia.org/w/index.php?title=Population_density *Contributors*: 1234r00t, 1297, 21655, 2mcm, 45Factoid44, 4halsop, A-giau, A. Parrot, A.Ou, Abhijitsathe, Achim Jäger, Addihockey10, Agentbla, Ahoerstemeier, Alansohn, AlexD, Alexjohnc3, Andre Engels, Andycjp, Anthonyz747, Aphaia, Aranel, Areology, Astuishin, Aude, Avicennasis, Aziz1005, B.d.mills, Barend, Barneca, Biederman, Blacklite, Blaircorrigan, Blanchardb, Bobblewik, Bogdangiusca, BrainyBabe, Brianlo747, Bricklayer, Brisvegas, BrokenSegue, Bsadowski1, Buchanan-Hermit, Bwilkins, CUSENZA Mario, Calabe1992, Candorwien, Catgut, Cdc, Cenarium, Chimpex, Ckatz, Cliff smith, Clovis Sangrail, Coffee, CommonsDelinker, Corpx, Courcelles, Cretog8, Cwgordon7, D. Cerv, DCEdwards1966, DL81292, DancingPenguin, Darth Panda, Davidcannon, Dekisugi, Desiphral, Diaboli, Docu, Donarreiskoffer, DoorFrame, Drmgrvy, Dysprosia, E Pluribus Anthony, ERobson, Eagle4000, EdGl, Edgar181, Edward, Eeyore tim, Egg Centric, Eggman64, Eleassar777, Elekhh, Elmsat, Elockid, Epbr123, Eric Sandholm, EugeneZelenko, Excirial, Extra999, Eyedubya, Farosdaughter, Figureskatingfan, Filemon, Func, GFW82, Garion96, Gcorral, GeeJo, Geremia, Gergis, Ghaly, Gidonb, Ginsengbomb, Glen, Gnosis1185, Gog, Gparalikidis, Graculus, Grafen, Gryffindor, HGB, HalfShadow, Hallows AG, Hamedog, Helix pomatia, HenryLi, Hermitage17, Heron, Hitman012, Hoplophile, HornetMike, Hsm rocks ya right, Hypertall, Hégésippe Cormier, INeedANameThatIsFunny, Ibbn, Introvert, Iridescent, Island, J.delanoy, JNW, JUSTM3-X, Jacob.jose, Jake Spooky, JavierMC, Jaw959, Jds10912, Jeff G., Joblio, JohnDoe0007, Johnuniq, Jorgeposada14, Joseph Philipsson, Joseph Solis in Australia, Josisb, Jpk, Jprksh, Jrein59, Kaare, Karl-Henner, Karok, Keenan Pepper, Keilana, KenFehling, Ketiltrout, Kgorman-ucb, Kingturkey, Klparrot, KnowledgeOfSelf, Kr4ft, Kurieeto, Kyle1278, L Kensington, Lakers, Latebird, LcawteHuggle, LeaveSleaves, Leks81, Leszek Jańczuk, Levineps, Liftarn, Lightmouse, Logose, Lotu, Lucas175, LukeStewart, Luxborealis, Lytsio, MER-C, MJCdetroit, MaNeMeBasat, Mack2, Malcolmxl5, MapsMan, Marc Lacoste, Marek69, Mark J, Martial75, Matt Yeager, Maximus Rex, McSly, Mgnbar, MidgleyDJ, Miguel.mateo, Mindmatrix, Mintleaf, Missmarple, Moreschi, Motoxrampage101, Mrbuddhafreak, Mschlindwein, Mxn, NLikhov, NTK, Nagy, Nakon, Naval Scene, Neko-chan, Nentrex, Nepenthes, NewEnglandYankee, Newbyguesses, Nick C, Noah Salzman, Nunquam Dormio, Nyttend, Olly150, Olsonnnnn, Ongar the World-Weary, Orestimba49, PBP, Panicburststs, Patrick, Pax:Vobiscum, Pearle, Pengo, Pgan002, Phaust, PhilKnight, Philip Trueman, Physchim62, Piano non troppo, PierreAbbat, Pigman, Pleasetouchme, Popsracer, Professionalsmartpeople1, Pumpmeup, Quasipalm, R Lowry, R'n'B, RLeopard, RTV User 545631548625, Radishes, Random User 937494, Readyhairman, Remuel, Rev. John, ULC, RexNL, Rgdboer, Riana, Richard Warrington, Rmhermen, Romanm, Ronhjones, RoyBoy, Ryaelt, Ryan Valdés, SDC, Samrager, SarekOfVulcan, Sarenne, Scohoust, Shizhao, Simmania, Sin Harvest, Skinnyweed, SmthManly, Snowolf, Snowolfd4, Spencer, Srasku, StaticGull, Ste, Steven Zhang, Steveprutz, Stormie, Sumahoy, Sunoco, Superm401, Surfbear335, Symode09, TGalt, Tabitcorne2, TaintedMustard, TakuyaMurata, Tangotango, Tanzy97, Taykp, Tbhotch, Tellyaddict, Terence, The Anome, The Hybrid, The Thing That Should Not Be, Thecaeser, Thehelpfulone, Thekohser, Thermalite, Thucydides of Thrace, Thunder8, Timeshifter, Tktktk, Tlc61, Tobias Galtieri, Tom harrison, Tomatoman180, Topbanana, Torchwoodwho, Tradewater, Tree Kittens, Trengarasu, Triesault, Tsxado, Tuanglen, Una Smith, Uncle Dick, Uuuussssseeeerrrrnnnnaaaammmmeeee, Vegard20, Violetriga, Vollex, Vranak, Waggers, Wangi, Wavelength, WeißNix, Wereon, Will Beback, Wilsbadkarma, Wimt, Yellowtruck, Ylem, Yngvadottir, Zanimum, Zap Rowsdower, Zfr, Zzuuzz, Александър, अीष भटनागर, 602 anonymous edits

Marriage *Source*: http://en.wikipedia.org/w/index.php?title=Marriage *Contributors*: ...adam..., 10metreh, 16@r, 200.191.188.xxx, 212.159.1.xxx, 2Wikid, 3eguoxn02, 4twenty42o, 5hin3, 66madewoqa, A8UDI, A;uoshefaqw832r3, ACSE, AV3000, Aaron Brenneman, Aaron Schulz, Abby724, Abeg92, Acather96, Acidburn24m, Adam Bishop, Adashiel, Adolf Diedrich, AdultSwim, Afaprof01, Agathman, Agh.niyya, Ahoerstemeier, Aitias, Akidd dublin, Akiko1092, Alan Liefting, Alansohn, Ale jrb, Alex contributing, Alex756, Alexdeburca18, Ali1986, Alienus, Alison, Alorlious, Alsandro, Altenmann, AnakngAraw, Anapple16, Anaraug, Anarchangel, And I feel like Alan Minter, Andrewlp1991, Andrewpmk, Andy Marchbanks, AndyKali, Andycjp, Andywatkins1888, Angela, Angelcastanyo, Angie Y., Angusmclellan, Aniten21, AnotherSolipsist, Antandrus, Anthony, Anthroprodigy, Apfox, Apoc2400, Arbiebrown, Arczi, Arichnad, Arjun01, Arkhunter, Armoreno10, Aroundthewayboy, Arteyu, Arthena, Artypants, Asarlaí, Aseel90, Ash211, Ashlie123, Aster77, Astronaught, Athkalani2000, Atlant, Atomaton, Attafei, Aurora, Awesomepeople11, AxelBoldt, Ayinyud, B Fizz, BD2412, BDock, BIGELLOW, BMF81, Baccyak4H, Bachrach44, Baconvitamins, Bajaboy, Bakanov, BalthCat, Bandit44, Bandros, Banes, Barneca, Barras, BarretBonden, Barrylb, Baseball Bugs, Basharh, Bazj, Bdean1963, Bdelisle, Beastiepaws, Beaucoup, Beechs, Bejinhan, Beland, Belchfire, Ben Standeven, Bennybp, Berek, Betacommand, Bewildebeast, Bhludzin, Bhuck, Biancasimone, BigDunc, BigFatBuddha, Birdmessenger, Black Jam Block, Blanchardb, Blargblarg09, Blondjamesblond, Bluemask, Bluntsword, Bmccleney, Bobo192, Bobri28, Bogdangiusca, Bogey97, Bomac, Bonadea, Bongwarrior, Bookandcoffee, Bookgrrl, Boothy443, Bornyesterday, Boycottthecaf, BozMo, BrainyBabe, Brian Boru, Brian Crawford, Brianga, BrokenSphere, Bronayur, Brooklynsam, BryanG, Bsayusd, Bsl235, Buddhainabucket, Burbanksalinger, Burschik, Bushcutter, Businessman332211, Butwhatdoiknow, C-w-l, C.Fred, CBDunkerson, CL, CLW, COMPFUNK2, CWii, Caesura, Caknuck, Calliopejen1, Caltas, CambridgeBayWeather, Camicakes, Can't sleep, clown will eat me, CanadianCaesar, CanadianLinuxUser, Canadiangurl, Canterbury Tail, Caper13, Capitan Obvio, Capricorn42, Captain Disdain, Carouselsinthesky, Catamorphism, Cathardic, CatherineMunro, Ccacsmss, Cenarium, Central2, CharlesHBennett, Charmed88, ChazBeckett, Cherubino, Chicocvenancio, Chonak, Chris 73, Chris the speller, ChrisCork, Chrism, Christian Guttmann, Christian75, Christopher Mann McKay, Christophorus, Cielomobile, Cirt, Ckatz, Cleared as filed, Cliché guevara, Clm17, Closedmouth, Clueless, Cmichael, Coelacan, Colincbn, Colonies Chris, CommonsDelinker, Conn1983, Consoar, Contaldo80, Conversion script, Coolcaesar, Cooljuno411, Coragma, Corpx, Correogsk, Cosmic Latte, Courcelles, CovenantD, Cparker, Cpl Syx, Cramertitan, Crateo12, CrazyLegsKC, Crumbsucker, Ctjf83, Cybercobra, CyborgTosser, CyclePat, Cyrius, Czolgolz, D'Agosta, DDerby, DJ Bungi, DJ Clayworth, DMSBel, DSRH, Da Joe, Da baum, Dabbler, Dadofsam, Daedalus969, Dainis, Damooney, Dan131m, DanB DanD, DanMS, DanielDeibler, Danikat, Dannyman, DarthSidious, Davelogan77, David Levy, David Shankbone, David91, DavidFarmbrough, Davodd, Dawud, Dayewalker, Dbachmann, Dcfleck, Deagle AP, Deanos, Decltype, Decorforyourhome, Deirdre Golash, Dekisugi, Delirium, Delldot, Deltabeignet, Denidowi, Denis tarasov, DerHexer, Derek Ross, Dethme0w, Dev920, Dhp1080, DickClarkMises, Diddi, Diego, Dina, Discospinster, Distad, Dixitque, Dkabecker, DocWatson42, Doldrums, Dominicano236, Domthedude001, Doniago, Donreed, Dowiha, Dpol, Drappel, Drew R. Smith, Drippingmintleaves, Drkamran, Drmies, Duffman, Dulla09, DuncanHill, Duoduoduo, Dybryd, E1e4n0r, EWS23, Eamonnca1, Ebzzzera39, Ec5618, Ed Poor, Edcolins, Edgar181, Edwardcowman, Eeekster, Egern, El C, ElTyrant, Elf, Elipongo, Emiao, Emmett5, Emmo827, Emojones, Empyrius, EncycloPetey, Englishnerd, Epbr123, Episiarch, EqualRights, Eric Forste, Erikp, Euchiasmus, Eukesh, Everyguy, Everyking, Evilandi, Ewawer, Excirial, Exir Kamalabadi, Eyu100, FF2010, FT2, Fabiform, Fagles, Fairlane75, FairmontMN, Fake522, Falcon8765, Falcon9x5, Fantrl, Fastily, Fdewaele, Fdsafdsa1, Ffdsajkl101, Fffdsajkl1000432, FilPines, Finisklin, Firsfron, Fishh, Fishhead64, Fishiehelper2, Fjeinca, Flamboyantcornhusker, Flammingo, Fleisher, Flewis, Flibjib8, Flickant, Flyers13, FocalPoint, Fplay, Franamax, Frank101, Frap, Fratrep, Frecklefoot, Fredrik, Frehley, FreplySpang, Fresheneesz, Frymaster, Fschmidt2, Fsiler, Fsushilcom, Funandtrvl, Future777, Fyrdbird, G0e St00PiD, GLGerman, GVnayR, Gabrielthursday, Gadfium, Gaius Cornelius, Galena11, Galoubet, Garhowell, Garik, Gastraphetes3, GateKeeper, Gdo01, Gdr, Gecko9999, Genius101, Geoff.green, Geometry guy, George100, Ghijak, Gianfranco, Gilliam, Gioueeoi, Glane23, Glenn, Gmaxwell, Gnangarra, Gobonobo, Godman123, Gogo Dodo, GoingBatty, Goldfritha, Gonzo fan2007, GoodDamon, Goodnightmush, Gorillasapiens, Gr8opinionater, GraemeL, Graham87, Greatgabriel, Greenguy1090, Gregbard, GregorB, Greudin, Griswaldo, Gsmgm, Guitarwizard39, Gurch, Gx872op, HGB, HJ Mitchell, Hadal, Haiduc, HalfDome, Hardyplants, Harryboyles, Hateless, Hayabusa future, Heard tried won, Heemz87, Helen Corby, Helicoptor, Henrygb, Hermesmessage, Hesitant contributor, Hibana, Hibernia345, Historyguy1965, Hmains, Homebuilding, Hoodinger10, Hooriaj, Hoshq, Hugh Mason, Husond, Hut 8.5, Hyacinth, IJA, IRP, Icarus3, IceCreamAntisocial, Iepeulas, Iisjarrod1, Iloveredhair, Imnotminkus, Indeterminate, IndulgentReader, Inferno, Lord of Penguins, Inkling, Insight08, InsufficientData, Inter, Interestedsister, Interwiki gl, Intrigue, Invader05, Irendraca, Iridescent, Iris91, Isaacsurh, Itohacs, Ixfd64, J Greb, J M Rice, J.delanoy, JASpencer, JBFrenchhorn, JForget, JNW, JPEsperanca&Nanda, JTSchreiber, Jaardon, Jackaranga, Jackfork, Jagged 85, Jake Wartenberg, JakeVortex, Jakes18, Jamescamry, Jamesmichaelsf, Jamesooders, JamieS93, Janus303, Janviermichelle, Jared Preston, Javert, Jay, Jayjg, Jdlh, Jeandré du Toit, Jeeves, Jeff G., Jeff Muscato, Jeff3000, Jeffpw, Jeffrey Mall, Jeph0112, Jesster79, JetLover, Jetzel73, Jfruh, Jgonzalez122, Jimothytrotter, Jimspilman, JinJian, Jiohdi, Jketola, Jmlk17, Jmueller0823, Jnestorius, Jobarts, Joe Decker, Joekoz451, Joetomjoe, John Kenneth Fisher, JohnC, Joie de Vivre, Jon513, Jonathunder, Jonstamos57, Jontomkittredge, Jooler, Jose77, Joseph Solis in Australia, Josephprymak, Joshaustintatemart, Joylifen, Joyous!, Jplatts, Jpritikin, Jrtayloriv, Jsc83, Jtdirl, Julesd, Jusdafax, Just zis Guy, you know?, Justin Eiler, Jvano, Jvb7988, Jwrosenzweig, JzG, Jzmonique, Jéské Couriano, K, K.d.stauffer, KJS77, KPH2293, Kaaveh Ahangar, Kablamo2007, Kafziel, Kaiba, Kansan, Kaobear, Karada, Karim S, Kasper4000, Katieh5584, Kauffner, Kaunietis30, Kc62301, Keegan, Kehrbykid, Kelovy, Kerowyn, Kevin B12, Kf4bdy, Khalid Mahmood, Khin2718, Khushi, Kiding, Kik-dawg, KillerChihuahua, Killerserv, KimvdLinde, King Lopez, King of Hearts, Kirill Lokshin, Kirti 1102, KissyKrissy1, Kitch, Kiyae, Klipper, Kmaguir1, Kmsiever, Kneale, KngLamoni, Knowledge-is-power, KnowledgeOfSelf, Knowpedia, Knucmo2, Korg, Kralljа, Krupo, Krótki, Ks0stm, Kukini, Kvng, LOL, Laca, LachlanA, Lacqui, LadyofShalott, Lagalag, Laikalynx, LambaJan, Laogeodritt, Lapsus Linguae, Last1in, Latka, Lauradimmock, Law, Lawrencekhoo, Leatherbud 287, LeaveSleaves, Lectonar, Lee J Haywood, LeeHunter, Leethepea08, Legokid, Lenoxus, Leon7, Lerdsuwa, Leszek Jańczuk, Liberal00Q1, Lightmouse, LilHelpa, Linestarz, Little guru, LittleOldMe, LizardJr8, Lloyd wallis2003, LogiJake, Lokantha, Loonymonkey, Lord Pistachio, Lotje, Lotti4tosh, Lquilter, Lugnuts, LuigiManiac, Luke w, Luna Santin, Lupin, MARussellPESE, MCTales, MK8, MPS, MPerel, MPorciusCato, MSGJ, MZMcBride, Machinkat, Mack2, Magister Mathematicae, Maiaminna, Mailer diablo, Makeemlighter, Malhonen, Malo, Mamalujo, Mandarax, Mangoe, Mangostar, ManiacK, Mannafredo, Manuel Trujillo Berges, Marcos, Mareino, Marek69, Margareta, Mark Renier, Mark Richards, Mark Ryan, Markb, Markruffolo, MarsRover, MartinHarper, Martinphi, Masaruemoto, Masterdojo, Materialscientist, Matt Deres, Matthew238, Mattkokes, Maunus, Maurice Carbonaro, Max Naylor, MaxHund, Mayooranathan, McSly, Mcbrooks, Mcmillin24, Mcy jerry, Mdev05, Mdraffi, Meanclevelandstreamingmachine, Media anthro, Mel Etitis, Melaen, Mentifisto, Mephistophelian, Metalfreak8153, Metamagician3000, Metatron's Cube, Methnor, Mhakcm, Michael Dinolfo, Michael Hardy, Michael Zimmermann, Michael2, Michaelgreenfeld, Midway, Mifter, Mikaey, Mike Rosoft, Minderbinder, Misaw, MishaPan, Misterx001, Mlouns, Mm35173, Montrealais, Morningstar2651, Mpbx3003, Mpondofranklin, Mpradeep, Mr Stephen, Mr. G. Williams, Mr.Z-man, MrBell, MrFish, MrOllie, Mrdthree, Mrmiscellanious, Mrzaius, Ms21608, Ms2ger, Mtcv, MuZemike, Mufka, Munin and hugin, Mushin, Mushroom, Mvjs, Mwazzap, Mxn, Myheadhurta, Mysdaao, Mythdon, N0nr3s, N5iln, NJMauthor, Nagy, Nakon, Nar12, Nard the Bard, NatGertler, Natl1, NawlinWiki, Ncmoulee, Nearfar, Neelix, Neilc, Neodanite, Neovu79, Nephron, Nereocystis, Netesq, Neutrality, Nev1, New4325, Newbyguesses, NewlywedsGuide, Newman Luke, Newone, Newsroom hierarchies, Newtman, Ngebendi, Nichetas, Nick Levinson, NickBush24, Nicke Lilltroll, Niew, Niex05, Nikai, Nikodemos, Nishkid64, Nkras, Nlu, No Guru, Noctibus, Non-Dairy Creamer, Normalguy818, Northgrove, Noted Seven, Nowhither, Ntennis, NuclearWarfare, Nuttycoconut, Nwbeeson, O, OMLOLRAWR, ONEder Boy, Oberlin, Occono, Octane, Oda Mari, OfForByThePeople, Ohmusama, Ohnoitsjamie, Ojigiri, Olegwiki, Oleykin, Oliverkeenan, Oliverkroll, Olivier, Omicronpersei8, Onorem, Orionus, Ortonmc, OscarMilde, Othniel Kenaz, Otr500, OverlordQ, Oxymoron83, PCHS-NJROTC, PUNKMOWEIRDO, Pablo2garcia, Paranomia, Paranormal Skeptic, Parthasarathy B, PassaMethod, Patrick, Patrickarzul, Paul Stansifer, Paulkramer, Pax:Vobiscum, Peculiar Light, Pedro, Peloneous, Penbat, People power, Perfect Proposal, Persian Poet Gal, Peruvianllama, PeshawarPat, Peter Deer, Petesmiles, Phanerozoic, Pharaoh of the Wizards, Pharohandy, Phil Sandifer, Philip Cross, Philip Trueman, PhilipTheArabian, Philippe, Phixt, Phunting, Phynicen, PiAndWhippedCream, PierreAbbat, PinchasC, Pinkadelica, Plasticup, Plrk, PocklingtonDan, Poindexter Propellerhead, Polly, Populus, Portress, PoshFan, Positiveones, Pranathi, Pre1mjr, Prjacoby, ProGloriaDei, Profero, Psy guy, Psychohistorian, Puellanivis, Pupster21, Putush, Pwningall, Pwningall2, Pyromania1967,

Pyromania1968, Python eggs, Q17, Qatter, Qscott86, Quadell, Quantpole, QueenCake, Queezbo, Querist, Quigley, Quintote, Qxz, R, R'n'B, R.123, RB1956, RHaworth, RK, RR, Ragazz, Rahm Kota, Rajkiandris, RakhiKapoor, Ramsankar2007, RandomP, Ranveig, RaseaC, Rashed, Ravichandar84, RayAYang, Rbj, Rdsmith4, Realm of Shadows, Recognizance, Reesh, Reinoutr, Reinyday, Renice, Retired username, Reverend Duck, Revived, RexNL, Rezd, Rfl, Rhobite, Rholton, Rich Farmbrough, Richard Arthur Norton (1958-), Richard001, Richard0612, Rickyrab, Riferimento, Rjakew, Rje, Rjwilmsi, Rm w a vu, Roadrunner, Robert Merkel, Robert. Helms, Robofish, RodnoseRave, Roland2, Ronels, Ronhjones, Ronz, Ropers, Roscelese, Rosemaryamey, Ross Fraser, Rosser1954, Rrburke, RussBlau, RxS, RyanEberhart, Ryulong, S007MI6, S3000, SD6-Agent, SEWilco, SQGibbon, SWAdair, Sa.vakilian, Sadads, Sadaf Fayyaz, Sadi Carnot, Saga City, SaintCahier, Salgueiro, Sam Blacketer, Sam Hocevar, Sam Korn, Sam RHCP, Sampi, Sanders muc, Sango123, Sannse, Sarahbeara126, Sardanaphalus, SasiSasi, Sauve.sean, Savidan, Sazbee, Sbrools, Sceptre, SchuminWeb, Scientizzle, Scjessey, Scohoust, Sdkfdslkf, Sdsds, Septegram, Seraphimblade, SeventhHell, Shadypalm88, Shalom Yechiel, Shanel, Shannon bohle, Shimirel, Shirik, Shizhao, Shoaler, Shward, Simesa, SimonP, Singhalawap, SiobhanHansa, Sir Trollsalot, Sjakkalle, Sketchesofskidrow, Sky Attacker, Skylark42, Slakr, Sleeping123, Sligocki, Slrubenstein, Slysplace, Smercer70, Smith2006, Smokizzy, Snavelz, Sneakalee, Snowolf, Someone65, Sonicrules2, Soylentyellow, SpaceFalcon2001, Spellcast, Spencerk, Splash, Spoileralerter, Srnec, Ssbohio, Ssundar, Sswonk, StaticGull, Stbalbach, Steel, Sten for the win, Stephen, Stephenb, SteveRamone, Stevekass, Steven J. Anderson, StewartNetAddict, Stezton, Stickee, Stone, Stopfollowingmedrapple, Stubblyhead, Student7, Stupid Corn, Subtorrct, Sujafu, SummerPhD, Supergeo, Superking, Supposed, Surfeited, Suvankarc, Swati.agrawal84, Symane, Syrthiss, THEN WHO WAS PHONE?, TXTXTXTXTX7, Tabletop, Tariqabjotu, Tasc, Tassedethe, TastyPoutine, Tazok, Tbhotch, Tcncv, Ted Longstaffe, Tediouspedant, TelusFielder, Tenuto123, Terryhi, TexasAndroid, Textorus, That Guy, From That Show!, Thaurisil, The Anome, The Duke of Waltham, The Man in Question, The Rambling Man, The Sartorialist, The Thing That Should Not Be, TheFix63, TheSun, TheTito, Theavengerofinformation, Thedeciderone, Theoneandonlyafghanlion, Therealhazel, Therefore, Thesituation212, Thingg, Thue, Tide rolls, TigerShark, Tim Ivorson, Timaeusj, Timc, Timrollpickering, Tkn20, Toad, Tobby72, Tocool4beci, Toffeebabe16, Tokek, Tom Lougheed, TomPhil, Tombomp, Tommy2010, Tomtheman5, Tomwsulcer, Tony1, Tonywalton, Touchdown Turnaround, Toussaint, Towsonu2003, Tpbradbury, Tracy49, Transcendence, Travelbird, Trevdna, Trishm, Trollderella, Truthspreader, Trystan, Tscribner, Tslocum, Twaz, Twoequalone, Tyler, Typhronemon5, UH Collegian, USTtommie09, Ubardak, Unbelievable999, Unclemikejb, Unschool, Used2BAnonymous, UserDoe, VKokielov, Valerie505, Valermos, Van helsing, Vanished user, Vedek Dukat, Vegaswikian, Velho, Vemblut, Verdesinistra, Versageek, VeryVerily, Vibup1, Vidkun, Vincent Gray, ViolaViera, Violetriga, Visorstuff, Vizcarra, VltavaHF, Volunteer Marek, VoodooKobra, Voortle, Vranak, Vsmith, Vssun, Vwilmot, WAS 4.250, WJBscribe, Wachowich, Warofdreams, WarthogDemon, Wavelength, Wavesmikey, Wayiran, Wedding planning ideas, Wejwfiawoiwe2kws3, WereSpielChequers, Whcernan, Where, White Shadows, Wiki Raja, Wiki-uk, WikiDao, WikiLaurent, WikidSmaht, Wikiklrsc, Wikitanvir, William Avery, WilliamSommerwerck, Wisco, Wizkid0607, Wjejskenewr, Wknight94, Woohookitty, WordyGirl90, WotherspoonSmith, Wpktsfs, Wsde, Wsvlqc, Wtmitchell, Wuzur, X!, X-factor, X42bn6, Xerxes247, Xgllo, XieChengnuo, Xp54321, Xspeed10trillion, Xullius, Yamaguchi, Yamamoto Ichiro, Yesyoudid, Yidisheryid, Youandme, Youngtoure88, Yuckfoo, YusufMJH, Yvwv, ZacharyS, Zahnrad, Zanfar, Zhou Yu, Zincsalt8199, Zocky, Zuejay, Zzuuzz, ارسان وبأ لي‌لو, 2260 anonymous edits

Poverty threshold *Source*: http://en.wikipedia.org/w/index.php?title=Poverty_threshold *Contributors*: Aaboltaabol, Acroterion, Aengus, Agradman, Ahackworth, Aillema, Aliyevramin, Arthena, Astronautics, AuburnPilot, Ausir, Barek, Beland, Bigpapi6292, Bluerasberry, Bookandcoffee, BrendelSignature, Burrburr, CLW, Can't sleep, clown will eat me, Cantus, Cardsplayer4life, Cassivs, Cconnett, Cflm001, Cielomobile, Ckatz, Classicrockfan42, Closenplay, Cntras, Cocoaguy, Coelhodg, Commander Keane, Common Man, Cretog8, Cuvtixo, Dambrosio, Damicatz, DanKeshet, David Biddulph, Dgies, Dismas, Dlohcierekim, Drmies, Duoduoduo, Dwrcan, Dystopos, Ed Poor, Edward, Egoadvocate, ElentariAchaea, ElinorD, Elitehobo, Epbr123, Eraserhead1, Eurosong, Fakelvis, Falconleaf, Fasttwitch, Fconaway, Fidalgo, Fitzaubrey, Foober, Frenzie, Gaff, Galf, Gavin.perch, Gene.arboit, Gilliam, Gizziiusa, Globalsolidarity, Goldom, Grick, Guoguo12, Hadrianheugh, Haoie, HarveyA, Hector.C.Jorge, Hede2000, Henrygb, Heron, HexaChord, Hibernian, Hrh80, Human.v2.0, Humanflux, Hurutch, Ian13, Ivan1984, Iviney, J.delanoy, JakubATusaDotcom, Janshouse, Javadane, JeremyA, Jhortman, Jimtaip, Jitse Niesen, Joseph Solis in Australia, Josh Parris, Joy, Junling, Kanook, KathrynLybarger, Kellen`, Kelly Martin, Ken@popp2.com, Khanra, Kirachinmoku, Kissmequick, KungFuMonkey, Lalit Jagannath, Leslie Mateus, Leszek Jańczuk, Levineps, LilHelpa, Logical Fuzz, Lost tourist, MOA12345, Maralia, Marek69, Mattbrundage, Maurreen, Maxis ftw, McDutchie, Metta Bubble, Million Moments, Minimac, Misanthr0pe, Monsday, Mrshannigan, Mshecket, Muhandes, Musiphil, Mydogategodshat, Mynameisnotpj, Mysdaao, Möchtegern, Nach0king, Nagy, Naval Scene, Neptunius, Netalarm, NigelR, Nihiltres, Nishkid64, Noozgroop, Oledr, Otolemur crassicaudatus, P.bertazzi, Paddles, Passionless, Philip Trueman, Phoebe, Piotrus, Policy Person, Postmortemjapan, Quest for Truth, R1000R1000, RJ010890, RainbowOfLight, Revth, RexNL, Riana, Rich Farmbrough, RobyWayne, Roke, Sam Hocevar, Sammo, Sapphic, Scarsxrunxdeep, SeanAhern, Search4Lancer, Septegram, Shannon bohle, Sharkface217, Sharnak, Sheetal.sehgal, Shoy, SiobhanHansa, Skysmith, Slhurst, Smallman12q, Smokecrack, Snipson23, Squashy, Strait, Tanaats, Team9a, Terjepetersen, The man stephen, Thingg, Tigga en, Timc, Timwi, Toh, Tommy2010, TouristPhilosopher, Twang, Ultramarine, Valdomiro, Volunteer Marek, WAGATHEE, WhatamIdoing, Wing Nut, Woutersmet, Yaktx, Yoyoyoitsrob, Zappa711, Zouf, 327 anonymous edits

Image Sources, Licenses and Contributors

Image: Illinois - background map.png *Source*: http://en.wikipedia.org/w/index.php?title=File:Illinois_-_background_map.png *License*: unknown *Contributors*: Caroig

Image: Illinois - outline map.svg *Source*: http://en.wikipedia.org/w/index.php?title=File:Illinois_-_outline_map.svg *License*: unknown *Contributors*: Caroig

File:Red pog.svg *Source*: http://en.wikipedia.org/w/index.php?title=File:Red_pog.svg *License*: unknown *Contributors*: Anomie

File:Map highlighting Antioch Township, Lake County, Illinois.svg *Source*: http://en.wikipedia.org/w/index.php?title=File:Map_highlighting_Antioch_Township,_Lake_County,_Illinois.svg *License*: unknown *Contributors*: User:Omnedon

File:Map of Illinois highlighting Lake County.svg *Source*: http://en.wikipedia.org/w/index.php?title=File:Map_of_Illinois_highlighting_Lake_County.svg *License*: unknown *Contributors*: User:Dbenbenn

File:Map of USA IL.svg *Source*: http://en.wikipedia.org/w/index.php?title=File:Map_of_USA_IL.svg *License*: unknown *Contributors*: Abnormaal, Hogweard, Huebi, Lokal Profil, Lupo, Mattbuck, Petr Dlouhý, 1 anonymous edits

File:Otthouse.jpg *Source*: http://en.wikipedia.org/w/index.php?title=File:Otthouse.jpg *License*: unknown *Contributors*: User:Shsilver

File:Lake County Illinois 1839.png *Source*: http://en.wikipedia.org/w/index.php?title=File:Lake_County_Illinois_1839.png *License*: unknown *Contributors*: User:Fishal

Image:USA Lake County, Illinois age pyramid.svg *Source*: http://en.wikipedia.org/w/index.php?title=File:USA_Lake_County,_Illinois_age_pyramid.svg *License*: unknown *Contributors*: user:WarX

File:Satellite image of Lake county.jpg *Source*: http://en.wikipedia.org/w/index.php?title=File:Satellite_image_of_Lake_county.jpg *License*: unknown *Contributors*: U.S. Geological Survey

Image:I-94.svg *Source*: http://en.wikipedia.org/w/index.php?title=File:I-94.svg *License*: unknown *Contributors*: Augiasstallputzer, Ltljltlj, SPUI, 1 anonymous edits

Image:US 12.svg *Source*: http://en.wikipedia.org/w/index.php?title=File:US_12.svg *License*: unknown *Contributors*: Bidgee, SPUI, 3 anonymous edits

Image:US 14.svg *Source*: http://en.wikipedia.org/w/index.php?title=File:US_14.svg *License*: unknown *Contributors*: Bidgee, SPUI, 3 anonymous edits

Image:US 41.svg *Source*: http://en.wikipedia.org/w/index.php?title=File:US_41.svg *License*: unknown *Contributors*: Rocket000, SPUI

Image:US 45.svg *Source*: http://en.wikipedia.org/w/index.php?title=File:US_45.svg *License*: unknown *Contributors*: Luigi Chiesa, SPUI, Xnatedawgx

Image:Illinois 21.svg *Source*: http://en.wikipedia.org/w/index.php?title=File:Illinois_21.svg *License*: unknown *Contributors*: SPUI

Image:Illinois 22.svg *Source*: http://en.wikipedia.org/w/index.php?title=File:Illinois_22.svg *License*: unknown *Contributors*: SPUI

Image:Illinois 43.svg *Source*: http://en.wikipedia.org/w/index.php?title=File:Illinois_43.svg *License*: unknown *Contributors*: SPUI

Image:Illinois 53.svg *Source*: http://en.wikipedia.org/w/index.php?title=File:Illinois_53.svg *License*: unknown *Contributors*: SPUI

Image:Illinois 59.svg *Source*: http://en.wikipedia.org/w/index.php?title=File:Illinois_59.svg *License*: unknown *Contributors*: SPUI

Image:Illinois 60.svg *Source*: http://en.wikipedia.org/w/index.php?title=File:Illinois_60.svg *License*: unknown *Contributors*: SPUI

Image:Illinois 83.svg *Source*: http://en.wikipedia.org/w/index.php?title=File:Illinois_83.svg *License*: unknown *Contributors*: SPUI

Image:Illinois 120.svg *Source*: http://en.wikipedia.org/w/index.php?title=File:Illinois_120.svg *License*: unknown *Contributors*: SPUI

Image:Illinois 131.svg *Source*: http://en.wikipedia.org/w/index.php?title=File:Illinois_131.svg *License*: unknown *Contributors*: Luigi Chiesa, SPUI

Image:Illinois 132.svg *Source*: http://en.wikipedia.org/w/index.php?title=File:Illinois_132.svg *License*: unknown *Contributors*: Luigi Chiesa, SPUI

Image:Illinois 137.svg *Source*: http://en.wikipedia.org/w/index.php?title=File:Illinois_137.svg *License*: unknown *Contributors*: SPUI

Image:Illinois 134.svg *Source*: http://en.wikipedia.org/w/index.php?title=File:Illinois_134.svg *License*: unknown *Contributors*: SPUI

Image:Illinois 173.svg *Source*: http://en.wikipedia.org/w/index.php?title=File:Illinois_173.svg *License*: unknown *Contributors*: Common Good, SPUI

Image:Map of Lake County Illinois showing townships.png *Source*: http://en.wikipedia.org/w/index.php?title=File:Map_of_Lake_County_Illinois_showing_townships.png *License*: unknown *Contributors*: User:TheCuriousGnome

Image:Raging Bull (Six Flags Great America).jpg *Source*: http://en.wikipedia.org/w/index.php?title=File:Raging_Bull_(Six_Flags_Great_America).jpg *License*: unknown *Contributors*: http://www.flickr.com/photos/hendrickspho tos/

File:Flag_of_the_United_States_(Pantone).svg *Source*: http://en.wikipedia.org/w/index.php?title=File:Flag_of_the_United_States_(Pantone).svg *License*: unknown *Contributors*: User:Dbenbenn, User:Indolences, User:Jacobolus, User:Technion, User:Zscout370

File:US-GreatSeal-Obverse.svg *Source*: http://en.wikipedia.org/w/index.php?title=File:US-GreatSeal-Obverse.svg *License*: unknown *Contributors*: U.S. Government

File:USA orthographic.svg *Source*: http://en.wikipedia.org/w/index.php?title=File:USA_orthographic.svg *License*: unknown *Contributors*: User:Addicted04

File:Increase2.svg *Source*: http://en.wikipedia.org/w/index.php?title=File:Increase2.svg *License*: unknown *Contributors*: Sarang

File:USATopographicalMap.jpg *Source*: http://en.wikipedia.org/w/index.php?title=File:USATopographicalMap.jpg *License*: unknown *Contributors*: U.S. Dept. of Commerce/National Climactic Data Center/NOAA Satellite and Information Service

File:Haliaeetus leucocephalus2.jpg *Source*: http://en.wikipedia.org/w/index.php?title=File:Haliaeetus_leucocephalus2.jpg *License*: unknown *Contributors*: Cacophony, Calliopejen, Clindberg, Dysmorodrepanis, Eusebius, Head, Jatkins, Kersti Nebelsiek, Kilom691, LX, Learnsales, MPF, Moebiusuibeom-en, R. Engelhardt, Richard001, Rrburke, Vanished user 001

Image:Map of USA with state names 2.svg *Source*: http://en.wikipedia.org/w/index.php?title=File:Map_of_USA_with_state_names_2.svg *License*: unknown *Contributors*: User:Andrew c

File:MayflowerHarbor.jpg *Source*: http://en.wikipedia.org/w/index.php?title=File:MayflowerHarbor.jpg *License*: unknown *Contributors*: William Halsall

File:Declaration independence.jpg *Source*: http://en.wikipedia.org/w/index.php?title=File:Declaration_independence.jpg *License*: unknown *Contributors*: Aavindraa, Amandajm, Beyond My Ken, Bohème, Cambalachero, Editor at Large, Elemaki, Fred J, Geni, GrawpSock, Hluup, Ibn Battuta, Mhby87, Misogi, Mutter Erde, Nonenmac, Panoptik, Patstuart, Pmlineditor, Razghandi, UpstateNYer, WTCA, Wst, Xavigivax, 16 anonymous edits

File:U.S. Territorial Acquisitions.png *Source*: http://en.wikipedia.org/w/index.php?title=File:U.S._Territorial_Acquisitions.png *License*: unknown *Contributors*: United States federal government (converted it from JPEG to PNG and retouched it)

File:Battle of Gettysburg, by Currier and Ives.png *Source*: http://en.wikipedia.org/w/index.php?title=File:Battle_of_Gettysburg,_by_Currier_and_Ives.png *License*: unknown *Contributors*: Beyond silence, Churchh, Kyro, Mtsmallwood, Möchtegern, PhilFree, Shizhao, Thuresson, UpstateNYer, 9 anonymous edits

File:Ellis island 1902.jpg *Source*: http://en.wikipedia.org/w/index.php?title=File:Ellis_island_1902.jpg *License*: unknown *Contributors*: AnRo0002, Aude, Infrogmation, Kjetil r, UpstateNYer, 8 anonymous edits

File:1944 NormandyLST.jpg *Source*: http://en.wikipedia.org/w/index.php?title=File:1944_NormandyLST.jpg *License*: unknown *Contributors*: Chief Photographer's Mate (CPHOM) w:Robert F. SargentRobert F. Sargent, U.S. Coast Guard

File:UA Flight 175 hits WTC south tower 9-11 edit.jpeg *Source*: http://en.wikipedia.org/w/index.php?title=File:UA_Flight_175_hits_WTC_south_tower_9-11_edit.jpeg *License*: unknown *Contributors*: User:UpstateNYer

File:Capitol Building Full View.jpg *Source*: http://en.wikipedia.org/w/index.php?title=File:Capitol_Building_Full_View.jpg *License*: unknown *Contributors*: Noclip

File:WhiteHouseSouthFacade.JPG *Source*: http://en.wikipedia.org/w/index.php?title=File:WhiteHouseSouthFacade.JPG *License*: unknown *Contributors*: User:UpstateNYer

File:USSupremeCourtWestFacade.JPG *Source*: http://en.wikipedia.org/w/index.php?title=File:USSupremeCourtWestFacade.JPG *License*: unknown *Contributors*: User:UpstateNYer

File:Hague Clinton May 14 2010 Crop.jpeg *Source*: http://en.wikipedia.org/w/index.php?title=File:Hague_Clinton_May_14_2010_Crop.jpeg *License*: unknown *Contributors*: User:UpstateNYer

File:USS Abraham Lincoln(CVN 72).jpg *Source*: http://en.wikipedia.org/w/index.php?title=File:USS_Abraham_Lincoln(CVN_72).jpg *License*: unknown *Contributors*: Photographer's Mate 3rd Class Jordon R. Beesley

File:Wall-Street Nueva York6397.JPG *Source*: http://en.wikipedia.org/w/index.php?title=File:Wall-Street_Nueva_York6397.JPG *License*: unknown *Contributors*: User:Poco a poco

File:Ranch style home in Salinas, California.JPG *Source*: http://en.wikipedia.org/w/index.php?title=File:Ranch_style_home_in_Salinas,_California.JPG *License*: unknown *Contributors*: BrendelSignature at en.wikipedia

File:Aldrin Apollo 11.jpg *Source*: http://en.wikipedia.org/w/index.php?title=File:Aldrin_Apollo_11.jpg *License*: unknown *Contributors*: NASA

File:Map of current Interstates.svg *Source*: http://en.wikipedia.org/w/index.php?title=File:Map_of_current_Interstates.svg *License*: unknown *Contributors*: User:SPUI

File:Harvard yard winter 2009j.JPG *Source*: http://en.wikipedia.org/w/index.php?title=File:Harvard_yard_winter_2009j.JPG *License*: unknown *Contributors*: chensiyuan

File:Texas Medical Center Aerial.jpg *Source*: http://en.wikipedia.org/w/index.php?title=File:Texas_Medical_Center_Aerial.jpg *License*: unknown *Contributors*: Hequals2henry

File:Census-2000-Data-Top-US-Ancestries-by-County-1396x955.png *Source*: http://en.wikipedia.org/w/index.php?title=File:Census-2000-Data-Top-US-Ancestries-by-County-1396x955.png *License*: unknown *Contributors*: User:Conscious, User:Louis Waweru

Image:NYC wideangle south from Top of the Rock.jpg *Source*: http://en.wikipedia.org/w/index.php?title=File:NYC_wideangle_south_from_Top_of_the_Rock.jpg *License*: unknown *Contributors*: User:Dschwen

File:LA Skyline Mountains2.jpg *Source*: http://en.wikipedia.org/w/index.php?title=File:LA_Skyline_Mountains2.jpg *License*: unknown *Contributors*: User:Nserrano

File:BrunswickWhiteChurch.jpg *Source*: http://en.wikipedia.org/w/index.php?title=File:BrunswickWhiteChurch.jpg *License*: unknown *Contributors*: User:UpstateNYer

File:Motherhood and apple pie.jpg *Source*: http://en.wikipedia.org/w/index.php?title=File:Motherhood_and_apple_pie.jpg *License*: unknown *Contributors*: Scott Bauer

File:PB050006.JPG *Source*: http://en.wikipedia.org/w/index.php?title=File:PB050006.JPG *License*: unknown *Contributors*: -

File:Kerouac by Palumbo.jpg *Source*: http://en.wikipedia.org/w/index.php?title=File:Kerouac_by_Palumbo.jpg *License*: unknown *Contributors*: Tom Palumbo from New York, NY, USA

File:Times Square 1-2.JPG *Source*: http://en.wikipedia.org/w/index.php?title=File:Times_Square_1-2.JPG *License*: unknown *Contributors*: User:UpstateNYer

File:Shea Smith-edit1.jpg *Source*: http://en.wikipedia.org/w/index.php?title=File:Shea_Smith-edit1.jpg *License*: unknown *Contributors*: User:Jjron

file:Census Bureau seal.svg *Source*: http://en.wikipedia.org/w/index.php?title=File:Census_Bureau_seal.svg *License*: unknown *Contributors*: United States Bureau of the CensusThis vector version by w:User:MysidMysid

file:US-CensusBureau-Logo.svg *Source*: http://en.wikipedia.org/w/index.php?title=File:US-CensusBureau-Logo.svg *License*: unknown *Contributors*: U.S. Government

File:U.S. Census Bureau Regions.svg *Source*: http://en.wikipedia.org/w/index.php?title=File:U.S._Census_Bureau_Regions.svg *License*: unknown *Contributors*: U.S. Department of Commerce Economics and Statistics Administration U.S. CENSUS BUREAU

File:Census Bureau headquarters, Suitland, Maryland, 2007.jpg *Source*: http://en.wikipedia.org/w/index.php?title=File:Census_Bureau_headquarters,_Suitland,_Maryland,_2007.jpg *License*: unknown *Contributors*: United States Census Bureau

Image:Countries by population density.svg *Source*: http://en.wikipedia.org/w/index.php?title=File:Countries_by_population_density.svg *License*: unknown *Contributors*: Miguel Contreras, Guatemala

File:World_population_density_1994.png *Source*: http://en.wikipedia.org/w/index.php?title=File:World_population_density_1994.png *License*: unknown *Contributors*: User:QuartierLatin1968

Image:Population density with key.png *Source*: http://en.wikipedia.org/w/index.php?title=File:Population_density_with_key.png *License*: unknown *Contributors*: User:Keenan Pepper, User:SG

Image:Crowd in HK.JPG *Source*: http://en.wikipedia.org/w/index.php?title=File:Crowd_in_HK.JPG *License*: unknown *Contributors*: User:Hamedog

File:TariatLandscape.jpg *Source*: http://en.wikipedia.org/w/index.php?title=File:TariatLandscape.jpg *License*: unknown *Contributors*: User:Doron

File:Wedding rings.jpg *Source*: http://en.wikipedia.org/w/index.php?title=File:Wedding_rings.jpg *License*: unknown *Contributors*: EvaK, FlickrLickr, FlickreviewR, Nevit

File:Brauysegen im Bett.gif *Source*: http://en.wikipedia.org/w/index.php?title=File:Brauysegen_im_Bett.gif *License*: unknown *Contributors*: AndreasPraefcke, Bohème, Esprit15d, Haabet, Wayiran, Xenophon, 2 anonymous edits

File:Lodewijk XIV-Marriage.jpg *Source*: http://en.wikipedia.org/w/index.php?title=File:Lodewijk_XIV-Marriage.jpg *License*: unknown *Contributors*: Antoine Dieu

File:Shinto married couple.jpg *Source*: http://en.wikipedia.org/w/index.php?title=File:Shinto_married_couple.jpg *License*: unknown *Contributors*: Aotake, Dodo, Melanom, Mikomaid, Reggaeman, Sakura-saku-kuni, Skipjack, Toto-tarou, Ytoyoda, 1 anonymous edits

File:Jan Brueghel the Elder-Christus und die Ehebrecherin-AP.JPG *Source*: http://en.wikipedia.org/w/index.php?title=File:Jan_Brueghel_the_Elder-Christus_und_die_Ehebrecherin-AP.JPG *License*: unknown *Contributors*: Anne97432, Finavon, Mattes, Shakko, TwoWings, Vincent Steenberg, Wst

File:V03p128a01 Ketubah.jpg *Source*: http://en.wikipedia.org/w/index.php?title=File:V03p128a01_Ketubah.jpg *License*: unknown *Contributors*: Faigl.ladislav, Malcolm Schosha, Red devil 666, Sheynhertz-Unbayg, Shmuel haBalshan, WhisperToMe

File:Rembrandt Harmensz. van Rijn 146.jpg *Source*: http://en.wikipedia.org/w/index.php?title=File:Rembrandt_Harmensz._van_Rijn_146.jpg *License*: unknown *Contributors*: A. Wagner, Anne97432, Bukk, Diomede, EDUCA33E, Emijrp, Foroa, Mattes

File:Israëls-A Jewish Wedding-1903.jpg *Source*: http://en.wikipedia.org/w/index.php?title=File:Israëls-A_Jewish_Wedding-1903.jpg *License*: unknown *Contributors*: Grendelkhan, Ilse@, Vincent Steenberg, 3 anonymous edits

File:St Mary's Church, Kyoto.jpg *Source*: http://en.wikipedia.org/w/index.php?title=File:St_Mary's_Church,_Kyoto.jpg *License*: unknown *Contributors*: Pmx

File:Phil and Marlene.jpg *Source*: http://en.wikipedia.org/w/index.php?title=File:Phil_and_Marlene.jpg *License*: unknown *Contributors*: Original uploader was Joshuajohanson at en.wikipedia

File:Nikah 003.jpg *Source*: http://en.wikipedia.org/w/index.php?title=File:Nikah_003.jpg *License*: unknown *Contributors*: Original uploader was Zainubrazvi at en.wikipedia

File:Muslim wedding in India.jpg *Source*: http://en.wikipedia.org/w/index.php?title=File:Muslim_wedding_in_India.jpg *License*: unknown *Contributors*: Claude Renault (Crop applied by uploader)

File:Hindu marriage ceremony offering.jpg *Source*: http://en.wikipedia.org/w/index.php?title=File:Hindu_marriage_ceremony_offering.jpg *License*: unknown *Contributors*: Jaisingh Rathore

File:Terryandmarkwedding.jpg *Source*: http://en.wikipedia.org/w/index.php?title=File:Terryandmarkwedding.jpg *License*: unknown *Contributors*: Ramsey2006, 1 anonymous edits

File:Thai Bride Price 2008.jpg *Source*: http://en.wikipedia.org/w/index.php?title=File:Thai_Bride_Price_2008.jpg *License*: unknown *Contributors*: Thomas & Joanna Ainscough. Original uploader was Tainscough at en.wikipedia

File:Percentage population living on less than $1.25 per day 2009.svg *Source*: http://en.wikipedia.org/w/index.php?title=File:Percentage_population_living_on_less_than_$1.25_per_day_2009.svg *License*: unknown *Contributors*: Tony0106

File:Percent poverty world map.png *Source*: http://en.wikipedia.org/w/index.php?title=File:Percent_poverty_world_map.png *License*: unknown *Contributors*: User:Sbw01f

GNU Free Documentation License Version 1.2, November 2002 Copyright (C) 2000,2001,2002 Free Software Foundation, Inc. 59 Temple Place, Suite 330, Boston, MA 02111-1307 USA Everyone is permitted to copy and distribute verbatim copies of this license document, but changing it is not allowed.

0. PREAMBLE

The purpose of this License is to make a manual, textbook, or other functional and useful document "free" in the sense of freedom: to assure everyone the effective freedom to copy and redistribute it, with or without modifying it, either commercially or noncommercially. Secondarily, this License preserves for the author and publisher a way to get credit for their work, while not being considered responsible for modifications made by others. This License is a kind of "copyleft", which means that derivative works of the document must themselves be free in the same sense. It complements the GNU General Public License, which is a copyleft license designed for free software. We have designed this License in order to use it for manuals for free software, because free software needs free documentation: a free program should come with manuals providing the same freedoms that the software does. But this License is not limited to software manuals; it can be used for any textual work, regardless of subject matter or whether it is published as a printed book. We recommend this License principally for works whose purpose is instruction or reference.

1. APPLICABILITY AND DEFINITIONS

This License applies to any manual or other work, in any medium, that contains a notice placed by the copyright holder saying it can be distributed under the terms of this License. Such a notice grants a world-wide, royalty-free license, unlimited in duration, to use that work under the conditions stated herein. The "Document", below, refers to any such manual or work. Any member of the public is a licensee, and is addressed as "you". You accept the license if you copy, modify or distribute the work in a way requiring permission under copyright law. A "Modified Version" of the Document means any work containing the Document or a portion of it, either copied verbatim, or with modifications and/or translated into another language. A "Secondary Section" is a named appendix or a front-matter section of the Document that deals exclusively with the relationship of the publishers or authors of the Document to the Document's overall subject (or to related matters) and contains nothing that could fall directly within that overall subject. (Thus, if the Document is in part a textbook of mathematics, a Secondary Section may not explain any mathematics.) The relationship could be a matter of historical connection with the subject or with related matters, or of legal, commercial, philosophical, ethical or political position regarding them. The "Invariant Sections" are certain Secondary Sections whose titles are designated, as being those of Invariant Sections, in the notice that says that the Document is released under this License. If a section does not fit the above definition of Secondary then it is not allowed to be designated as Invariant. The Document may contain zero Invariant Sections. If the Document does not identify any Invariant Sections then there are none. The "Cover Texts" are certain short passages of text that are listed, as Front-Cover Texts or Back-Cover Texts, in the notice that says that the Document is released under this License. A Front-Cover Text may be at most 5 words, and a Back-Cover Text may be at most 25 words. A "Transparent" copy of the Document means a machine-readable copy, represented in a format whose specification is available to the general public, that is suitable for revising the document straightforwardly with generic text editors or (for images composed of pixels) generic paint programs or (for drawings) some widely available drawing editor, and that is suitable for input to text formatters or for automatic translation to a variety of formats suitable for input to text formatters. A copy made in an otherwise Transparent file format whose markup, or absence of markup, has been arranged to thwart or discourage subsequent modification by readers is not Transparent. An image format is not Transparent if used for any substantial amount of text. A copy that is not "Transparent" is called "Opaque". Examples of suitable formats for Transparent copies include plain ASCII without markup, Texinfo input format, LaTeX input format, SGML or XML using a publicly available DTD, and standard-conforming simple HTML, PostScript or PDF designed for human modification. Examples of transparent image formats include PNG, XCF and JPG. Opaque formats include proprietary formats that can be read and edited only by proprietary word processors, SGML or XML for which the DTD and/or processing tools are not generally available, and the machine-generated HTML, PostScript or PDF produced by some word processors for output purposes only. The "Title Page" means, for a printed book, the title page itself, plus such following pages as are needed to hold, legibly, the material this License requires to appear in the title page. For works in formats which do not have any title page as such, "Title Page" means the text near the most prominent appearance of the work's title, preceding the beginning of the body of the text. A section "Entitled XYZ" means a named subunit of the Document whose title either is precisely XYZ or contains XYZ in parentheses following text that translates XYZ in another language. (Here XYZ stands for a specific section name mentioned below, such as "Acknowledgements", "Dedications", "Endorsements", or "History".) To "Preserve the Title" of such a section when you modify the Document means that it remains a section "Entitled XYZ" according to this definition. The Document may include Warranty Disclaimers next to the notice which states that this License applies to the Document. These Warranty Disclaimers are considered to be included by reference in this License, but only as regards disclaiming warranties: any other implication that these Warranty Disclaimers may have is void and has no effect on the meaning of this License.

2. VERBATIM COPYING

You may copy and distribute the Document in any medium, either commercially or noncommercially, provided that this License, the copyright notices, and the license notice saying this License applies to the Document are reproduced in all copies, and that you add no other conditions whatsoever to those of this License. You may not use technical measures to obstruct or control the reading or further copying of the copies you make or distribute. However, you may accept compensation in exchange for copies. If you distribute a large enough number of copies you must also follow the conditions in section 3. You may also lend copies, under the same conditions stated above, and you may publicly display copies.

3. COPYING IN QUANTITY

If you publish printed copies (or copies in media that commonly have printed covers) of the Document, numbering more than 100, and the Document's license notice requires Cover Texts, you must enclose the copies in covers that carry, clearly and legibly, all these Cover Texts: Front-Cover Texts on the front cover, and Back-Cover Texts on the back cover. Both covers must also clearly and legibly identify you as the publisher of these copies. The front cover must present the full title with all words of the title equally prominent and visible. You may add other material on the covers in addition. Copying with changes limited to the covers, as long as they preserve the title of the Document and satisfy these conditions, can be treated as verbatim copying in other respects. If the required texts for either cover are too voluminous to fit legibly, you should put the first ones listed (as many as fit reasonably) on the actual cover, and continue the rest onto adjacent pages. If you publish or distribute Opaque copies of the Document numbering more than 100, you must either include a machine-readable Transparent copy along with each Opaque copy, or state in or with each Opaque copy a computer-network location from which the general network-using public has access to download using public-standard network protocols a complete Transparent copy of the Document, free of added material. If you use the latter option, you must take reasonably prudent steps, when you begin distribution of Opaque copies in quantity, to ensure that this Transparent copy will remain thus accessible at the stated location until at least one year after the last time you distribute an Opaque copy (directly or through your agents or retailers) of that edition to the public. It is requested, but not required, that you contact the authors of the Document well before redistributing any large number of copies, to give them a chance to provide you with an updated version of the Document.

4. MODIFICATIONS

You may copy and distribute a Modified Version of the Document under the conditions of sections 2 and 3 above, provided that you release the Modified Version under precisely this License, with the Modified Version filling the role of the Document, thus licensing distribution and modification of the Modified Version to whoever possesses a copy of it. In addition, you must do these things in the Modified Version: A. Use in the Title Page (and on the covers, if any) a title distinct from that of the Document, and from those of previous versions (which should, if there were any, be listed in the History section of the Document). You may use the same title as a previous version if the original publisher of that version gives permission. B. List on the Title Page, as authors, one or more persons or entities responsible for authorship of the modifications in the Modified Version, together with at least five of the principal authors of the Document (all of its principal authors, if it has fewer than five), unless they release you from this requirement. C. State on the Title page the name of the publisher of the Modified Version, as the publisher. D. Preserve all the copyright notices of the Document. E. Add an appropriate copyright notice for your modifications adjacent to the other copyright notices. F. Include, immediately after the copyright notices, a license notice giving the public permission to use the Modified Version under the terms of this License, in the form shown in the Addendum below. G. Preserve in that license notice the full lists of Invariant Sections and required Cover Texts given in the Document's license notice. H. Include an unaltered copy of this License. I. Preserve the section Entitled "History", Preserve its Title, and add to it an item stating at least the title, year, new authors, and publisher of the Modified Version as given on the Title Page. If there is no section Entitled "History" in the Document, create one stating the title, year, authors, and publisher of the Document as given on its Title Page, then add an item describing the Modified Version as stated in the previous sentence. J. Preserve the network location, if any, given in the Document for public access to a Transparent copy of the Document, and likewise the network locations given in the Document for previous versions it was based on. These may be placed in the "History" section. You may omit a network location for a work that was published at least four years before the Document itself, or if the original publisher of the version it refers to gives permission. K. For any section Entitled "Acknowledgements" or "Dedications", Preserve the Title of the section, and preserve in the section all the substance and tone of each of the contributor acknowledgements and/or dedications given therein. L. Preserve all the Invariant Sections of the Document, unaltered in their text and in their titles. Section numbers or the equivalent are not considered part of the section titles. M. Delete any section Entitled "Endorsements". Such a section may not be included in the Modified Version. N. Do not retitle any existing section to be Entitled "Endorsements" or to conflict in title with any Invariant Section. O. Preserve any Warranty Disclaimers. If the Modified Version includes new front-matter sections or appendices that qualify as Secondary Sections and contain no material copied from the Document, you may at your option designate some or all of these sections as invariant. To do this, add their titles to the list of Invariant Sections in the Modified Version's license notice. These titles must be distinct from any other section titles. You may add a section Entitled "Endorsements", provided it contains nothing but endorsements of your Modified Version by various parties--for example, statements of peer review or that the text has been approved by an organization as the authoritative definition of a standard. You may add a passage of up to five words as a Front-Cover Text, and a passage of up to 25 words as a Back-Cover Text, to the end of the list of Cover Texts in the Modified Version. Only one passage of Front-Cover Text and one of Back-Cover Text may be added by (or through arrangements made by) any one entity. If the Document already includes a cover text for the same cover, previously added by you or by arrangement made by the same entity you are acting on behalf of, you may not add another; but you may replace the old one, on explicit permission from the previous publisher that added the old one. The author(s) and publisher(s) of the Document do not by this License give permission to use their names for publicity for or to assert or imply endorsement of any Modified Version.

5. COMBINING DOCUMENTS

You may combine the Document with other documents released under this License, under the terms defined in section 4 above for modified versions, provided that you include in the combination all of the Invariant Sections of all of the original documents, unmodified, and list them all as Invariant Sections of your combined work in its license notice, and that you preserve all their Warranty Disclaimers. The combined work need only contain one copy of this License, and multiple identical Invariant Sections may be replaced with a single copy. If there are multiple Invariant Sections with the same name but different contents, make the title of each such section unique by adding at the end of it, in parentheses, the name of the original author or publisher of that section if known, or else a unique number. Make the same adjustment to the section titles in the list of Invariant Sections in the license notice of the combined work. In the combination, you must combine any sections Entitled "History" in the various original documents, forming one section Entitled "History"; likewise combine any sections Entitled "Acknowledgements", and any sections Entitled "Dedications". You must delete all sections Entitled "Endorsements".

6. COLLECTIONS OF DOCUMENTS

You may make a collection consisting of the Document and other documents released under this License, and replace the individual copies of this License in the various documents with a single copy that is included in the collection, provided that you follow the rules of this License for verbatim copying of each of the documents in all other respects. You may extract a single document from such a collection, and distribute it individually under this License, provided you insert a copy of this License into the extracted document, and follow this License in all other respects regarding verbatim copying of that document.

7. AGGREGATION WITH INDEPENDENT WORKS

A compilation of the Document or its derivatives with other separate and independent documents or works, in or on a volume of a storage or distribution medium, is called an "aggregate" if the copyright resulting from the compilation is not used to limit the legal rights of the compilation's users beyond what the individual works permit. When the Document is included in an aggregate, this License does not apply to the other works in the aggregate which are not themselves derivative works of the Document. If the Cover Text requirement of section 3 is applicable to these copies of the Document, then if the Document is less than one half of the entire aggregate, the Document's Cover Texts may be placed on covers that bracket the Document within the aggregate, or the electronic equivalent of covers if the Document is in electronic form. Otherwise they must appear on printed covers that bracket the whole aggregate.

8. TRANSLATION

Translation is considered a kind of modification, so you may distribute translations of the Document under the terms of section 4. Replacing Invariant Sections with translations requires special permission from their copyright holders, but you may include translations of some or all Invariant Sections in addition to the original versions of these Invariant Sections. You may include a translation of this License, and all the license notices in the Document, and any Warranty Disclaimers, provided that you also include the original English version of this License and the original versions of those notices and disclaimers. In case of a disagreement between the translation and the original version of this License or a notice or disclaimer, the original version will prevail. If a section in the Document is Entitled "Acknowledgements", "Dedications", or "History", the requirement (section 4) to Preserve its Title (section 1) will typically require changing the actual title.

9. TERMINATION

You may not copy, modify, sublicense, or distribute the Document except as expressly provided for under this License. Any other attempt to copy, modify, sublicense or distribute the Document is void, and will automatically terminate your rights under this License. However, parties who have received copies, or rights, from you under this License will not have their licenses terminated so long as such parties remain in full compliance.

10. FUTURE REVISIONS OF THIS LICENSE

The Free Software Foundation may publish new, revised versions of the GNU Free Documentation License from time to time. Such new versions will be similar in spirit to the present version, but may differ in detail to address new problems or concerns. See http://www.gnu.org/copyleft/. Each version of the License is given a distinguishing version number. If the Document specifies that a particular numbered version of this License "or any later version" applies to it, you have the option of following the terms and conditions either of that specified version or of any later version that has been published (not as a draft) by the Free Software Foundation. If the Document does not specify a version number of this License, you may choose any version ever published (not as a draft) by the Free Software Foundation. ADDENDUM: How to use this License for your documents To use this License in a document you have written, include a copy of the License in the document and put the following copyright and license notices just after the title page: Copyright (c) YEAR YOUR NAME. Permission is granted to copy, distribute and/or modify this document under the terms of the GNU Free Documentation License, Version 1.2 or any later version published by the Free Software Foundation; with no Invariant Sections, no Front-Cover Texts, and no Back-Cover Texts. A copy of the license is included in the section entitled "GNU Free Documentation License". If you have Invariant Sections, Front-Cover Texts and Back-Cover Texts, replace the "with...Texts." line with this: with the Invariant Sections being LIST THEIR TITLES, with the Front-Cover Texts being LIST, and with the Back-Cover Texts being LIST. If you have Invariant Sections without Cover Texts, or some other combination of the three, merge those two alternatives to suit the situation. If your document contains nontrivial examples of program code, we recommend releasing these examples in parallel under your choice of free software license, such as the GNU General Public License, to permit their use in free software.

MIX
Papier aus verantwortungsvollen Quellen
Paper from responsible sources
FSC® C105338

Printed by Books on Demand GmbH, Norderstedt / Germany